Mathematics Problem Posing in Action

Shuk-kwan S. Leung

Mathematics Problem Posing in Action

Students' Learning, Teachers' Professional Development, and Parental Involvement

Springer

Shuk-kwan S. Leung
Institute of Education
National Sun Yat-sen University
Kaohsiung, Taiwan

ISBN 978-981-96-8479-3 ISBN 978-981-96-8480-9 (eBook)
https://doi.org/10.1007/978-981-96-8480-9

This Springer imprint is published by the registered company Springer Nature Singapore Pte Ltd.
The registered company address is: 152 Beach Road, #21-01/04 Gateway East, Singapore 189721, Singapore

Foreword by Kaye Stacey

I am delighted to be asked to write a foreword to this account of the work of Shuk-kwan (Susan) Leung. For me, the book makes two main contributions. First, it is an account of research and experiences of using problem posing as part of mathematics teaching. Second, it is a case study of how a teacher educator can organise her academic responsibilities and activities to make a significant impact on the practice of mathematics teaching in schools.

Problem posing is an important part of the everyday work of mathematics teachers. This book sets out the reasons why problem posing (here always closely related to the topics being taught) can also be a valuable part of the students' experience and how that might be achieved. The central goals for students are metacognitive: to help students see the underlying mathematical structure of the problems they are learning to solve (rather than only their surface features) and also to alert them to problem components that significantly affect difficulty. Problem posing is thus an example of a 'role reversal' teaching strategy—in this case, where students take on what is normally the task of their teacher.

The chapters are soundly based on the international research literature and embedded in practice. They show in a practical way how creating good problem posing tasks depends on teachers' pedagogical content knowledge for teaching and how this knowledge can usefully be extended to students in an age-appropriate way. All the studies draw on research findings about children's solution methods and difficulties, and the variety of problem types that are commonly encountered and need to be included in instruction. The three sections are each completed by an in-depth commentary from an international scholar who is well known for work on problem posing. These sections provide a broader perspective for Leung's work, as well as highlighting key findings and methods that will be helpful to other researchers and teacher educators.

Sitting behind these accounts is the story of how Shuk-kwan Leung has ensured that her long career in teacher education has not been an 'ivory tower' exercise but has made an impact on students' learning of mathematics. This requires her adopting many different roles, and throughout the book she uses an interesting literary device to highlight the multitude of tasks she is undertaking by calling herself Susan, Professor

Leung, Ms Leung and Mytle. In this way we see the effort that is required for teacher educators to make an impact on practice and the rewards that it can deliver. As Shuk-kwan explains at the end of Chap. 1, this book invites readers to accompany a band of co-learners as they walk through a time tunnel on a wonderful and fruitful journey.

April 2025

Kaye Stacey
University of Melbourne
Melbourne, VIC, Australia

Foreword by John Mason

Mathematics as a noun is for me the world of recorded mathematical thinking to be found in libraries and on the internet: a collection of problems, resolutions of and conjectures about those problems, and ways of thinking about and working on those problems, what Karl Popper referred to as a 'third world' (Popper, 1965). *Mathematical Thinking* as a gerund is for me a way of being in the material world, asking questions, appreciating structural relationships, justifying conjectures, and being stimulated to enter a world of imagination in which those structural relationships exist. A major challenge is to express insights about those structural relationships in gestures, diagrams, and symbols, thereby providing access to them from the material world.

Although solving problems is for me the stuff of mathematics, it has been my experience that the posing of problems, the search for explanations, and the recognition of structural relationships is where the energy comes from. It involves a transfer of initiative from teacher/parent to learner. As Bill Brookes suggested (1976; see also Brookes, 1966; Mason, 2016) the key question is not *what is a problem, but when is (something) a problem?*. His, and my answer is, when someone takes it on *as* a problem, in other words, takes the initiative.

All this is to say that for me, problem posing has always been an essential part of living a life informed by mathematical thinking. It is with great pleasure, therefore, to encounter this description of how Shuk-kwan (Susan) Leung has been integrating curriculum-relevant and curriculum-focused problem posing in her work with teachers and students. Careful distinctions are made between various forms of problem posing, from random, open problems, to curriculum-relevant problems. But the point is for students (and teachers) to experience not-knowing at the same time as feeling an urge to find out. The reader is led through the forest of constructs which constitute the relevant mathematics education literature, and shown how it can be called upon practically to provide students with an (even 'the') essential ingredient for learning to think mathematically, namely, problem posing.

Leung's account is delivered in three sections, focused respectively on Students' Learning, Teachers' Professional Growth, and Parental Involvement, reflecting transformations in her comprehension of the multiple factors at play in what, on the surface, appears to be a simple matter of ‘teaching mathematics’. Each section contains specific case studies highlighting transformations in her appreciation of various forces at play, and a commentary by an internationally respected contributor which places her work in context, and emphasises the effectiveness and breadth of reach of her contributions.

At the heart of this inspiring account of how her action research developed and enriched her appreciation of the complexities inherent in learning to think mathematically is the ways in which she sought to establish productive working relationships with people playing different roles in the lives of students, namely both teachers and parents, and in the background, the institutions through which these encounters take place. Leung’s enterprise is to show that mathematical problem posing plays an essential role in the teaching and learning of mathematics, and that pedagogically it is entirely possible and within the powers of teachers, once they have experienced it for themselves, and once they have appreciated how learning to pose problems for oneself can transform student engagement.

April 2025

John Mason
Open University
Milton Keynes, UK

University of Oxford
Oxford, UK

References

Brookes, W. (Ed.). (1966). *The development of mathematical activity in children: The place of the problem in this development.* ATM, Nelson.

Brookes, W. (1976). Philosophy and action in education: When is a problem? *ATM Supplement, 19*, 11–13.

Mason, J. (2016). When is a problem? “When” is actually the problem! In P. Felmer, E. Pehkonen, & J. Kilpatrick (Eds.), *Posing and solving mathematical problems: Advances and new perspectives* (pp. 263–285). Springer.

Popper, K. (1965 revised).*Conjectures and refutations: The growth of scientific knowledge.* Routledge & Kegan Paul.

Preface

Mathematical problem posing (MPP) has been a meaningful activity to me in many ways along my career path. When I taught as a secondary school teacher in Hong Kong, I assigned students to do MPP while waiting for friends to finish, but I did not use it for all students. When I pursued my doctoral studies in Pittsburgh, MPP was the heart of my dissertation and was part of my research agenda after graduation. After being a teacher educator in Taiwan, I introduced MPP and learned about it with students, teachers, and parents. Sharing what I gained in this journey is most delightful. To me, this sharing is a responsibility.

This book was conceived at a very special moment. I had the idea of writing it when the ICME 14 invitation arrived, and also when the online invited lecture was incompletely presented during the pandemic. The audience responded positively to my lecture, and their questions urged me to continue sharing the story in a book with a broader audience. The review of the proposal returned constructive comments and suggested sharing with practitioners as well. I added content-grade-specific illustrations in each chapter and made sure they did not overlap with each other. Although research assures the value in problem posing, it is not often seen in practice. One day, I told myself, "*I must try MPP and do it systematically*". A teacher educator tries something new before he or she introduces it to others. A caretaker tries new food before feeding a baby. This book shows that it is feasible to practice MPP not just for once but for different grade levels and many topics over time. I also drew a curriculum map to show which grade a certain topic is given in each chapter (see Fig. 3 of Chap. 1). However, MPP is not a panacea. One must also attend to context, semantics, representation, and even choice of numbers, for *content*, *participants*, and *situation.* The activity might be structured or not creative, but when a situation calls, then open MPP tasks are used. One reviewer noted that the book can be used to teach courses for prospective teachers. During the revisions, I reminded myself constantly to keep a balance between research and practice. The birth of this book was not smooth, but no matter what, I was determined to finish this book.

The cases were on students learning MPP, teachers teaching MPP, and parents involved in MPP. Data came from my funded research projects, graduate thesis advisees, teacher-training sessions, district supervision, and parents' workshops.

An applause goes to the conscientious participants (students, teachers, and parents). I am so lucky to be able to learn with them. If the circles in a Venn diagram represent types of participants in my project, then we can share knowledge only when the circles lie on the same plane. We were equal partners who walked hand in hand through this time tunnel. As I looked back, this book reports how bringing people together can be accomplished using a simple yet powerful act of MPP. This activity can be done anywhere. It creates opportunities for curiosity and connections. It makes mathematics meaningful and shared. MPP helps us to see mathematics not as a boring subject to be learned in school, but as a playful way to think, to wonder, and to grow together.

The name of this book is "Mathematics Problem Posing in Actions".

Who took the actions?

What kinds of actions were they?

How could that happen?

I leave the readers to imagine and guess the answers to the above questions. An overview is given in Chap. 1. Please read it first. Finally, I invite readers to enjoy reading it as much as I enjoyed writing it, if not more.

Kaohsiung, Taiwan
April 2025

Shuk-kwan S. Leung

Acknowledgments The completion of this book is a challenging and unforgettable experience. I am indebted to five respected scholars. Three contributed commentaries: Lyn English, Edward Silver, and Marta Civil. Lyn's publications often add knowledge to the state-of-the-art on thinking and learning. Ed's influential projects on teachers' professional development drive me to get research into practice. Upon reading Marta's research with immigrant parents, I followed in her footsteps and studied with parents. Two returned forewords, Kaye Stacey and John Mason, to enlighten the author and provide the readers with a mathematical-thinking portrait. Their words captured my hidden message in the book and supported the final production of the book. Knowing these five role models along my academic path is a great fortune. Silver is my dissertation advisor; meeting him change my path to academia. I thank him for being my academic parent.

I admire a group of creative and hardworking teachers and graduate students who completed data collection and initial analysis, and agreed to let me take charge of the final data analysis and manuscript: Jih-Pu Chang, Pei-Chi Chen, Kun-Chuan Chao, Chin-Chien ChuangYang, Yun-Qing Huang, Hsien-Chung Kuo, Tzu-Ling Kuo, Cheng-Ju Lee, Chai-Hui Wu, Hui-Wen Wu, and Hui-Ju Yang (in alphabetical order of surnames). I still remember when, where, and how we had meetings. I equally admire the efforts of anonymous parents who took time off to share knowledge with me. They joined me to witness the children's growth with pride. The teachers and parents still get together after the projects. Furthermore, I appreciated the assistance in the administration and data collection process: Yi-Chia Sun, Vesta Han, Pei-Jou Wu, Pei-Chi Wu, Kokoro Pan, Chun-Hsiung Lin, and Chia-Huang Hsu.

I am lucky to have a team that reads drafts and discusses the presentation of analysis and curricular issues. They are: Chia-Huang Chen, Iuan-Jyh Wang, Huan-wen Cheng, Elena Stoyanova, Jin-Hui Tang, Jen-Chieh Ho and Yi-An Cho. There was also a cheerful team for assistance in editing and formatting: Vicky Wang, Bowie Mok, Sandy Wu, Tiffany Hsu, and, Jean Wu. I also thank those who gave input, yet forgive me if I miss their names here.

A vote of thanks goes to the funding agency, the National Science and Technology Council (NSTC, formerly MOST, NSC) in Taiwan, for funding my research projects on problem solving and problem posing. I also thank Springer for its services in producing this book. A deep gratitude is delivered to three anonymous reviewers for the proposal and another three for the whole book, both for specific comments (e.g., page-wise questions or suggestions) and general comments (e.g., relationships among chapters). All of them provoked thoughts for vigorous revisions. I also thank my institutions for nice work environments (workspace, library services, and staff) and scenic campuses: National Chiayi University (Taiwan), The Chinese University of Hong Kong, and National Sun Yat-sen University (Taiwan).

The never-ending support comes from my family members. They gave me the power to face challenges and carry a smile at work. My husband, Chun-Kong, helps me to make a decision, yet does not decide for me. He makes sure that there is always a person to listen to me. My two gentlemen children, Pok-Man and Po-Yun, have never stopped challenging me since their birth. Teaching mathematics lessons in their elementary school was unforgettable to all of us in the family. Finally, I cannot thank our four adorable parents enough. They are my teachers whose stories in raising us confirmed the importance of parental engagement.

I now send in my appreciation to each of the above who cares. *Thank you.*

Kaohsiung, Taiwan
April 2025

Shuk-kwan S. Leung

Competing Interests The author has no competing interests to declare that are relevant to the content of this manuscript.

Contents

Chapter 1
Mathematical Problem Posing in Action: A Journey from Students' Learning, Teachers' Professional Growth to Parental Involvement

1 Mathematical Problem Posing: One Construct, Many Manifestations

Mathematics problem posing (MPP) is one construct. In a well-cited paper, problem posing is the formulation of new problems or the reformulation of problems into novel problems (Silver, 1994). In this activity, the problem poser is a provider of information (Simon, 1973), and, creates problems by providing a structure from own perspective (English, 2003). The voyage of problem solving includes problem posing, which leads to mathematics in the making (see Polya, 1945). However, when a person poses by creating a new problem, the action is different from a second kind of posing problems, which means presenting a known problem differently, or, a third kind of posing problems, which means making up test items. The three kinds of posing problems correspond to three translations into Chinese characters.

擬題 (nǐ tí): Constructing and formulation of a new problem (Brown & Walter, 1983; Silver, 1994).

佈題 (bù tí): Presenting a problem properly or posing problems differently to meet instructional objectives (Butts, 1980; Leung, 1991).

命題 (mìng tí): Making up test items for examinations (Leung, 1996).

Although the above three kinds of problem posing result in the birth of a problem, the three activities are different, and, distinctions need to be made.

The *nǐ tí* activity is for the formulation of a new problem, meaning from no problem to a problem; or from one known problem to a novel problem. The person who poses problems learns some mathematics after finding solutions. The other two, *bù tí* and *mìng tí*, are often for instructional and assessment purposes. Teachers or test developers make efforts to better present problems for target persons to solve. The three kinds of problem posing are related, yet different, in terms of purposes.

The author of this book is a mathematics teacher educator (MTE, see Sect. 4.1 for more details about the author). Over 30 years in her career, the author has worked on

S. S. Leung, *Mathematics Problem Posing in Action*,
https://doi.org/10.1007/978-981-96-8480-9_1

projects including all three kinds of problem posing (*nǐ tí, bù tí and mìng tí*). However, her studies were mostly on *nǐ tí*. The problem is posed by a person who solves it himself/herself or hopefully finds someone who can solve it. Unless otherwise stated, MPP means *nǐ tí* from now.

What is typical of *nǐ tí*?

Leung (2013) discussed four characteristics. First, MPP can be idiosyncratic. When one is considering some givens and poses a problem, one makes connections from givens to a goal according to how one "sees". Suppose the given is, "There are 15 boys and 20 girls in a class." a person may ask, "How many children are there altogether?" but another person may ask, "What is the ratio of boys to girls?" Each person "sees" a problem that may be different from what others "see". Second, the act of problem posing involves plausible reasoning (e.g. "Consider the change in this ratio if the total number of children is the same but the number of boys is 11 instead?"). Third, problem posing can happen before, during, or after problem solving. For example, after posing and solving for the total number of children, one may ask, "What is the percentage of boys in the class?" Fourth, a posed problem that is not yet solved can be insufficiently specified, impossible, or even impossible, such as "If 25 bottles were distributed to 15 boys and 20 girls and each one gets one, how many bottles were left?".

In addition to the four characteristics, problem posing is often considered an inseparable part of problem solving.

According to Polya (1945), there are four phases in problem solving: (1) understanding the problem, (2) devising a plan, (3) carrying out the plan, and, (4) looking back. However, a straightforward procedure of passing from the first to the fourth phase may not work. If there is a difficulty, the problem solver's going back and forth is common. Each switch to a different phase signals a person's evaluation of what he or she did in the current phase. When a step is not successful the solver will backtrack, which is represented by counter-clockwise arrows. And, if a person is solving the problem posed by oneself, rather than a problem that has been given by others, the initial Understand phase is considered a Pose phase, and, the solver understands the problem posed by oneself. Subsequently, problem posing naturally occurs before, during, or after solving a problem (given or posed by oneself). The decisions and actions of posing and solving can be inter-related (Fig. 1).

Indeed, problem posing and problem solving are inseparable. The cycle "Pose-Plan-Carry Out-Look Back" explains the never-ending process of a person posing

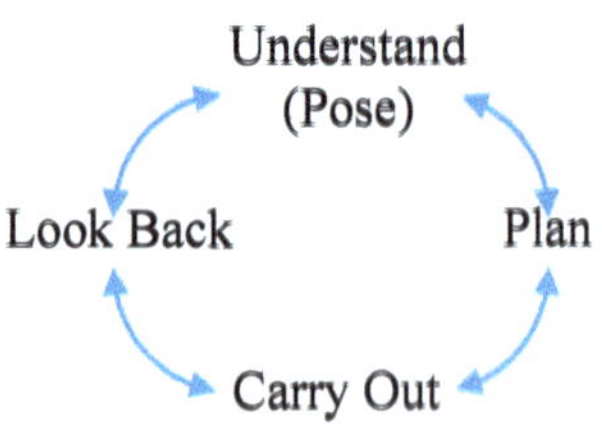

Fig. 1 Four phases in problem posing and problem solving

and problem solving (Leung, 2009). It is customary for one to pose a problem and then solve it (Pose-then-solve); and, it is also natural for one to solve a problem and then pose a problem (Solve-then-pose).

Indeed, problem posing has always been an important direction for educational research in mathematics, and also in chemistry and physics (Ellerton & Ellerton, 1987; Mestre, 2002). The importance of problem solving and problem posing in mathematics education is highlighted in well-known books (Brown & Walter, 1983; Freudenthal, 1973; Polya, 1954) and handbooks on mathematics education research (Cai, 2017; Grouws, 1992; Lester, 2007). Problem posing reports also appeared in focus issues in well-known journals such as *Educational Studies in Mathematics, International Journal of Educational Research, ZDM-The International Journal on Mathematics Education*, and, *International Journal for Mathematics and Science Education* or as forums in international meetings on mathematics education (e.g., ICME, PME).

Problem posing in mathematics is also important in educational practice. "*Whether as a goal itself or as a means to other goals, mathematical problem posing appears to have taken root both in the educational practice community and in the research community*" (Silver, 2013, p. 161). In teaching, MPP can be done individually or in groups, carrying psychological or social values. Problem-posing education, coined by Paulo Freire in his 1970 book *Pedagogy of the Oppressed*, is a method of teaching that emphasizes critical thinking for liberation. Freire calls traditional pedagogy the "*Banking model of education*". He argues that pedagogy should instead treat the learner as a co-creator of knowledge (Wikipedia). Indeed, in doing MPP, a student is a co-creator of math knowledge.

Whether in research or practice, there was attention to categories for problem-posing situations and structures. Reitman (1965) considered four cases of problem situations; in each case, MPP can be performed. The combinations of two specifications (defined or undefined) and two states (given and the goal), result in the four cases (R1, R2, R3, R4). Besides these four cases, MPP has two meanings (Silver, 1994): the reformulation of known problems (see left-two column under Silver S1) or the formulation of new problems (S2). Silver's S1 included posing problems properly (Butts, 1980) or making test items (Leung, 1996). In addition, MPP can be classified according to structure (see Stoyanova & Ellerton, 1996): structured (SE1), semi-structured (SE2), or free (SE3). Finally, in a professional development project for teachers, Leung (2013) introduced 6 categories of instructional MPP tasks. In this inventory by Leung (2013), the categories were named according to their given state: A known problem (solve-then-pose); an Algorithm, a Text, a Figure or Table, a Math Topic, or an Answer (pose a problem when solved will yield the given answer). The Table 1 explains the relationship between Reitman, Silver, Stoyanova & Ellerton, and Leung.

Note that Leung (2013) did not include completely free posing, which means, just posing any problem. The reason is that Leung (2013) was focusing on tasks that were conveniently used by elementary school teachers in an implementation project on math problem posing. In free posing, the aims were sparse and the results were hard to explain and use for instruction.

Table 1 Categories of problem posing according to given and goal

Reitman (1965)	Silver (1994)	Stoyanova and Ellerton (1996)	Leung (2013)
Given defined Goal defined (R1)	Reformulate known problems (S1)	Structured (SE1)	L1. Known problem
Given defined Goal undefined (R2)	Formulate new Problems (S2)	Semi-structured (SE2)	L2. Algorithm L3. Text L4. Figure/Table L5. Math Topic
Given undefined Goal defined (R3)	Formulate new Problems (S2)	Semi-structured (SE2)	L6. Answer
Given undefined Goal undefined (R4)	Formulate new Problems (S2)	Free (SE3)	

Mathematical problem posing is considered as a product or process. *Product*. The above table of problem posing specifications represented by an MPP task is based on whether the given or the goal is defined, or to the extent to which it is defined. However, the MPP task can be further divided into two types according to the final product of problem posing: (a) posing a mathematics problem using a situation and (b) posing a story problem matching a fixed piece of mathematics. For example, the problem-posing activity asking one to pose one easy and one difficult percent problem (see van den Heuvel-Panhuizen et al., 1995) belongs to the former type of product. An example is "What is 10% of 50?" (a difficult problem). On the other hand, posing word problems matching a given division algorithm ($12 \div 8$; answer $= 13$; $12 \div 8$; answer $= 12$; $12 \div 8$; answer $= 12.5$) belongs to the latter type of product. An example is "Grandfather gives his grandchildren a box containing 80 balloons, which they share equally. How many balloons does each grandchild get?"

Process. Problem posing has also been classified according to processes. In a study conducted in Cyprus, four MPP processes were considered: Editing, Selecting, Comprehending, and Translating. This classification was used to study quantitatively how the four processes are related to children's performance (Christou et al., 2005). However, it is unclear if classifying tasks by process produces mutually exclusive categories. For example, in doing MPP, can a person exhibit only one process and not the other three? Say, can a person do only editing and not selecting, comprehending, or translating? Other than research on the classification of processes, there was an attempt to document problem posing processes. Leung (1994) created a method of using GPS behavior graphs to explain MPP processes (how one posed a sequence of problems). In her study, the problem poser used answers to previously posed problems as given to pose subsequent problems; the graphs showed the linkages among the problems.

To start, I say MPP is a construct, and after going through the above I continued to say, "It carries many manifestations" which was also the word given in a worthy paper by Silver (2013). In a classic paper by Kilpatrick (2004, also in 1978) he listed various variables in research on mathematics problem solving. Three major independent variables are *task, subjects, and situations.* I follow with the three variables below, as they are also important variables for research in math problem posing.

2 Task Variable in Research on Mathematics Problem Posing

2.1 Math Tasks as Variables in Math Problem Posing

A math task is important in affecting how students learn and convey messages about "what doing mathematics entails" (NCTM, 1991, p. 24). In a book on Soviet study on determining children's math ability, Krutetskii (1976) used three categories of tasks about information given: *information gathering, information processing, and information retention.* However, the same task, the problem posed depends on how the subject works on it. In *information gathering*, one task was, "A pupil bought 2b notebooks at one store and 3.5 times as many at another." (Krutetskii, 1976, p. 106). One student may pose a problem like "How many notebooks did he buy altogether". Another student might ask, "If b is 5, how many did he buy in the first store and the other store?" If in a group/class setting when the teacher or peers share possible problems, the "same task" would appear to be different in richness. Krutetskii (1976) designed a total of 79 tests of tasks, packed under 26 series (problems or systems of problems) that can be divided into 8 groups of ability. Students had opportunities to do *information gathering, information processing, and information retention* through problem posing and problem solving.

Stein, Smith, Henningsen, and Silver (2010) distinguished 4 types of tasks: tasks that require only *memorization*, tasks asking for *routine without connections*, tasks *with connections,* and tasks for *doing math.* The book included a variety of tasks and cases of enactment from the QUASAR project, and the examples given are helpful to schools for professional training of math teachers. However, whether a task is for memorization or doing math is related to the amount and process of gathering information, processing information, and information retention (Krutetskii, 1976).

As given above, one cannot tell what problem posing is, until one fully understands the category of its task, what is given, what is required to take action, and what product is required (a story situation matching a given math structure or a math structure matching a context), and the process taken…etc. The results of using tasks for problem posing also depend on who (the subject) and how (the situation). This is because the task provides affordance for the person involved in the activity (Schoenfeld, 2024).

2.2 Tasks for Different Subjects

One cannot comment if a task is good or bad, but one can tell if a task is appropriate or not for *Whom*. Polya (1945), in the book "*How To Solve It*", considered tasks as good if they were proportionate to the ability level of students. "Rich math tasks for whom be just right......are neither too hard nor too easy." In hill climbing, when the hiker cannot climb a mountain then the height of the mountain may not be proportionate to the ability level of hikers. Indeed, whether "solvers are incapable" or "tasks too hard" are indistinguishable. Thus, the correct proportion of challenge in a task cannot be judged until we consider the ability of the students.

Task Variables in Mathematical Problem Solving is a book that is broad and comprehensive, and, includes decisions on systematic manipulation of tasks for certain instructional purposes (Goldin & McClintock, 1979). It is a great resource for teacher educators and teachers, no matter what type of tasks, types of research on tasks, or instructions on problem solving using the tasks. In this book, there was one chapter by Kulm, "The Classification of Problem Solving Research Variables." It said, "The development of problem-solving ability is a cumulative process which depends on the history of problem-solving experience of the student. What is crucial to any problem-solving experience is the task itself. To advance knowledge about problem-solving, it is thus important that close attention be given to the characteristics of problem tasks" (Goldin & McClintock, 1979). The task or collection of tasks is the measuring instrument that is used to study the phenomena of problem solving. An understanding of how the variables describing the task itself interact with the total situation is a basic requirement for such a study. The ability to classify and define task variables would make it possible to control them systematically, to determine their effects on problem-solving behavior. The term "task variable" will mean any characteristic of problem tasks that assume a particular value. The study of task variables also has important implications for classroom instruction.

2.3 Tasks and Subjects for Different Situations

One cannot comment on a task design for *Who* is proper or not unless the situation for use is known. Kilpatrick (1978) wrote, "Any study of problem solving in mathematics involves a person (*subject*) solving a mathematical problem (*task*) under some condition (*situation*) (Kilpatrick, 1978)". The three in parenthesis were independent variables. The task was further classified into sub-categories like content, context, structure, and format. Subject variables included sub-categories like organismic (e.g. age, gender) and trait (e.g. attitude, interests). Situation variables included sub-categories (complete task in class or at home, what time of the day, how the teacher enacts the task). People often think that the situation does not matter. However, Kilpatrick remarked, "Unfortunately, they may." The situation does matter. Therefore, in a parallel branch of the tree called mathematics problem posing, the situation stands and it is as important as other branches.

As above, task, subject, and situation are also three independent variables in research on problem posing. Situations refer not only to the environment on individual solving but also classroom factors. Henningsen and Stein (1997) investigated classroom factors that may hinder or support students' thinking in mathematics; those factors were variables in situations. The mathematics task framework (MTF) was built. The MTE included tasks as given in *curricular materials, as set up by the teacher*, and, *as implemented by students*, and how the three affected *students' learning outcomes.* Hsu and Silver (2014) addressed curricular materials; they found that carefully sequencing a set of tasks made a difference in teaching geometry calculation tasks. Boston and Smith (2009) attended to how teachers set up tasks and found that increasing the cognitive demands of the tasks changed the ways tasks were implemented by students. Given the above, it is worthwhile for investigators to design a study by addressing the task, subject, and, situation. Also, to study factors for in-class problem posing and, to document students' learning outcomes.

The above two sections highlight the importance of MPP in problem-solving research (1) and prior work on task variables (2). Research along these two lines was geared towards learning and teaching, cognitive and affective domains, and constituted a basis for including different participants or co-learners in research and practices.

3 Subject Variables: Teachers and Parents as Co-Learners

Here, teachers mean math teacher educators, teachers, and prospective teachers. This book attends to different types of subjects (participants) and they worked and learned together. The investigator, a math teacher educator, is an action researcher herself. One of her continual efforts since 1993, is to implement research results on mathematical problem posing (MPP) into math instruction in the schools in Taiwan. In initial studies on MPP, she studied products and processes of prospective teachers in the United States and Taiwan (Leung, 1993, 1994, 1997) and found that prospective teachers were inexperienced in MPP but that they expressed interest in this activity and showed improvement after they did problem posing themselves. Later, she found that the evaluation method on posed problems was badly needed. Merely finding that teachers could not evaluate problem posing is not enough. The next step is to show how and give them specific example tasks and methods of analysis and assessment. Another effort she made was to develop a systematic assessment (or evaluation) method for MPP (Leung & Silver, 1997). Though there are no standard criteria for evaluating a problem, an analysis of posed problems can hint at what posers see in a given situation, what they consider as mathematical, and allow inferences about what they "do know" or "don't know".

In Taiwan, she introduced MPP to college students. She offered a two-credit course on problem posing for prospective elementary school teachers, using various research-based MPP tasks to familiarize students with MPP over 18 weeks.

Next, to introduce MPP to in-service teachers, the teacher educator incorporated MPP into invited talks for teacher-training workshops at national, state, or district levels. When there was an imbalance in the number of invited talks from states in the North, Central, South, and East, she actively made contact with those institutions that did not send her invitations and suggested giving a talk on MPP. She found the translations of problem posing into Chinese (*nǐ tí, bù tí and mìng tí*) helped teachers to distinguish one kind of MPP from the other two kinds of MPP. Hence, she tried to introduce problem posing to practitioners all over the country. By and by, a list of graduate theses on problem posing followed. In graduate programs with practitioners' enrolment, she offered graduate seminars on problem-solving and extended her efforts on problem posing to graduate students who are school teachers or prospective school teachers. When graduate students expressed difficulties in tasks given in readings she included in-class activities and used teaching aids. She found that understanding the tasks and solving them would help them to comprehend the readings. The actions and results of her investigations were reported by Leung (2017).

She also worked with Wu, a veteran elementary school teacher, and published papers on problem posing (Leung & Wu, 1999, 2000). Leung and Wu (1999) explained how MPP could happen when accidentally a problematic problem existed. Leung and Wu (2000) reported how parents and children posed problems at home and communicated findings to the teacher by diary writing.

In the year 2021, she delivered an invited lecture at ICME 14, Shanghai. The one-hour talk allowed her to share her attempts to use MPP as a means to study learning, teaching, and parenting. In this lecture, Leung explained how she walked through her journey and asked three questions: How do students learn via problem posing? How to change teacher education so that teachers can teach problem posing? How can problem posing empower parents to get involved in children's learning?

I. ***"How do students learn via problem posing?"***

To constitute one's knowledge, one must be curious. When curious one raises questions (pose) and finds answers (solve). After getting answers, one may ask more questions (pose more problems). This cycle goes on and on. (Fig. 2) Math competence is reached when one "poses and answers questions in or using mathematics" (Niss & Højgaard, 2019).

She made arrangements and taught children MPP. When children first attempted problem posing, they did not know what they were supposed to do and asked her questions such as, "Which problem(s) should I pose?", "Can you give me examples?", "Who is going to solve it?", and, "What if I cannot find the answer?" She came across difficulties in teaching. Therefore, she decided to work with teachers and research on how to teach children MPP, for different math contents, and, for various grade levels.

Fig. 2 Posing and solving

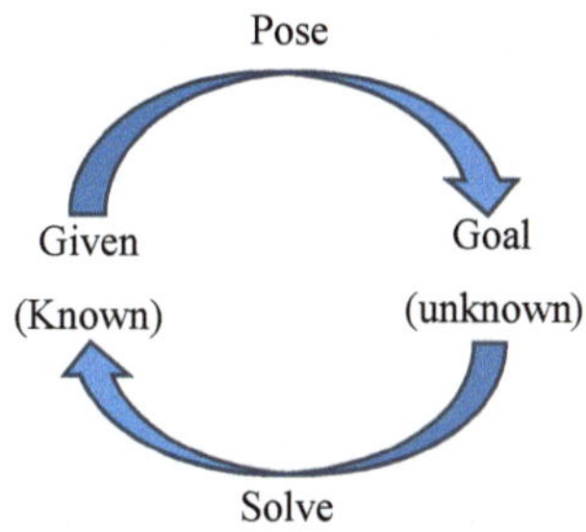

II. ***"How to change teacher education so that teachers can teach problem posing?"***

She believed that the know-how on teaching MPP depended on math content. She borrowed classes from school teachers and let them observe how she taught content-specific MPP. This was only a start. Interestingly, the teachers who observed also asked her questions. They might ask questions like, "Where can I find MPP tasks?", "How often shall I do MPP?", or "What can I do after children pose problems?" A larger project, on the integration of MPP, was conducted with groups of teachers in south Taiwan. The outcomes included resource books with MPP tasks and typical types of students' responses. She also used results to prepare prospective teachers. In all, she used her position as supervisor (at the national level) to bridge district supervisors, in-service teachers, and prospective teachers. Some conscientious teachers joined her in action research. In all, there were cases on different math contents, and for students at different grade levels. Some teachers expressed that parents asked questions about MPP when they were supervising children at home. The teacher needed help on how to work with parents. Parental involvement was her next step.

III. ***"How can problem posing empower parents to get involved in children's learning?"***

The math teacher educator continued to walk on this journey. The third part was unexpected, and *not all* parents' activities were on problem posing. It is just one of the tasks in parents' workshops. To start, a simple written work was clipped together with children's homework. The MPP homework was optional. Parents and students completed written work on MPP (grade 1 to grade 3). The returned rates and responses suggested that when written MPP tasks related to school math were ready, parents were willing to be involved. Also, she supervised a grade four school teacher to invite all parents to do MPP at home. Finally, she offered Friday afternoons family math camps for parents and children (K to grade four) and for parents (grade five, grade six). Dissemination of results were in books and on her website. Although there were different activities designed for parents and family members over the years, she will just focus on MPP.

In all, the one-hour invited lecture for ICME 14 was delivered via on-line, with highlights on how she answered the above three questions (on students, teachers, and parents as participants). The contents of the invited lecture were extended and organized into this book.

4 Situation: Roles and Actions of an MTE in and Out of the Ivory Tower:

This section augments the prior sections on MPP, *task* variables, and *subject* variables; and, explains the Taiwan context that provides opportunities for the author, a math teacher educator (MTE), to work with teachers and parents under waves of math curriculum during her career.

4.1 About the Author Who Is an MTE

In "*Those Who Understand: Knowledge Growth in Teaching*" (Shulman, 1986), Shulman started with the saying "*Those who cannot, teach!*". He developed the speech to include teaching as a profession and the importance of pedagogical content knowledge. Towards the end, he replaced the saying to, "*Those who understand, teach!*". In the year 2015, the author of this book had the chance to meet this scholar in person. Shulman gave a talk in Taipei and continued to stress the importance of studying teaching. He told a story to describe researchers, "A drunker lost something in the dark yet searched for it under the street lamp and not in the dark." He meant that researchers carried out studies in convenient settings. Shulman said, "*That's what we do so often. We look where there are lights but not on what we need to understand.*" He highly recommended researchers and practitioners to enter the classroom and study teaching, sometimes breaking the rules, as that may lead to surprises.

In the recall of the author, after teaching in a secondary school for nine years, she earned a doctoral degree in math education and taught in a teachers' college that prepared elementary school teachers. Her experience in teaching elementary was minimal. She gave her first lecture to elementary school teachers on math problem posing. The elementary school teachers were not listening to her. On the stage and from the floor, she heard, "*Those who do not teach anymore, teach the teachers*!"

This was a shock back in 1993, her first talk as a teacher educator.

The voice from the floor reminded her of Shulman (1986). "*Those who understand, teach!*" and also the drunker who "*look where there are lights*" instead of the dark.

> *I was that drunker too, I tended to study prospective teachers at the university. I never went to the schools to study in-service teachers. I did not try teaching elementary school children. I had no experience in teaching math problem posing.*

She found herself to be a professor hiding inside the Ivory Tower. She read books and suggested teachers teach according to the books. She decided to step out of the Ivory Tower. She decided to go back to school to teach.

But I am a professor now. How can I do that?

Let me take action.

First, she used teaching aids and taught pre-service teachers as if they were schoolchildren. Second, she taught children math as a parent during homeroom period. Third, she borrowed classes to teach in teaching training sessions. Finally, she conducted family math camps for parents. She looked for chances to teach MPP from kindergarten to grade 12. In short, she went to school; she practiced what the books suggested before she introduced herself to teachers and parents. She did that when Taiwan was undergoing curriculum reform.

4.2 A Bit of Taiwan Context: Curriculum Reform

During the past 30 years, Taiwan has undergone several waves of curriculum reform and was in between globalization and localization (Yang et al., 2022). For example, the year 1993 marked the change in having students actively construct math knowledge and not acquiring it through teachers' transmission. This was influenced by Western theories on constructivism. Also, the standards (MOE, 1993) included an emphasis on problem solving, communication, reasoning, and connections, which was influenced by standards in the US (NCTM, 1989). The year 2001 announced another wave called Nine Year Curriculum Alignment which included a remarkable change in naming after grouping or re-grouping of subject areas. Mathematics, however, was neither divided nor merged with another subject and stayed as one single subject area as before–Mathematics. The document lined up unified standards from grades one through nine. In the year 2014, the Twelve-year Basic Education reform document for *all* subjects was announced (MOE, 2014). Committees were then set up to work on standards for each of the five subject areas. The mathematics curriculum standard was completed (MOE, 2018). By 2019, compulsory education was on the way.

Although Taiwan is a small island on the world map, its name in math learning and math teaching was recognized. In 2007, Taiwan first participated in TIMSS and its math performance was second to Finland (the first among 59 regions). The wave 1993 constructivism was a student-centered approach and an affective element was to have them share their views generously and accept views from others. The recognition is not only on math learning but also on math teaching. Taiwanese elementary and high school teachers performed well on assessments specific to both mathematical content and mathematical pedagogical content knowledge and there were comprehensive and well-designed teacher education programs for pre-service and in-service teachers (Yang et al., 2022). A study on preservice math teacher training from 17 regions, ranked Taiwan as number one, when the focus was on the relationship between teacher preparation policies and the quality of future teachers, using multiple measures in mathematics content and pedagogy content knowledge (see Ingvarson & Rowley, 2017).

In recent curriculum standards, documents included math problem posing explicitly. The Nine-Year Curriculum Alignment document suggests to check if children understand a given expression, teachers can invite them to pose a real-life problem matching the expression (MOE, 2008). In the Twelve-Year Basic Education curriculum standards, math problem posing can be found in the curriculum guidebook for mathematics (MOE, 2020, February, p. 81). On p. 41, grade 1 or grade 2 children pose a word problem matching a given addition or subtraction algorithm $5 + 3 = 8$, $8–3 = 5$.

Students. The current system in education. Elementary school for 6 years (age 7–12), middle school for 3 years (age 13–15), senior high for 3 years (age 16–18), college for 4 years (age 19–23). The current system offers 12 years of compulsory education (including grade 1 to grade 12, but not kindergarten). *Teachers.* Teacher training and requirements is a related degree (e.g. math) with an additional 26 education credits for middle and high school levels. College with an additional 40 education credits for elementary school teachers (teach nearly all subjects). Professional development sessions for in-service secondary math teachers are arranged on one afternoon (depending on districts) when all math teachers do not have to teach. For elementary teachers, they do not teach on all Wednesday afternoons (to attend professional development sessions). *Parents.* They are involved in their children's learning at home and in school. For elementary school students, morning study sessions (when teachers have staff meetings) are conducted by morning mothers (volunteers who are caring parents). Although not all, some parents are volunteers in schools, to spend time learning how to supervise their children's homework and accompany them, or even learn together with teachers and children. Parents need directions and are resources in enhancing children's math learning.

5 Overview of the Book

The book consists of actions taken in trying mathematics problem posing (MPP) in Taiwan. It is a sincere sharing from a mathematics teacher educator (MTE) who taught students and worked conscientiously with school teachers and parents over the 30 years of her career. The actions were taken as she assumed different roles in a research agenda on MPP. At the same time, the actions were influenced by the country's aims in curriculum change in ways of teaching being advocated by official documents. In retrospect, this MTE referred to frequently asked questions, and related stories and reflected on the actions taken place over the time tunnel. She shared her learning in finding answers to questions on *Who, What, and How.*

Who was involved? [Subject variable, or, participant variable]

What piece(s) of mathematics needs to be learned and how can it be learned/taught? [Task variable]
How problem-posing/solving activities were intentionally used in actions? [Situation variable]

The three questions correspond to *subject, task, and situation* variables (and methodologies) in research on problem solving identified by Kilpatrick (1978). It makes sense to apply them in work on problem posing. First, the *subjects* or *participants* were students from kindergarten to high school, teachers from pre-service, in-service to veteran teachers, and, parents from families with diverse backgrounds. Second, the chapters reported on *tasks* given in math contents such as addition/subtraction, four algorithms, rational numbers, ratio and proportion, time and length, symmetry, surface areas of solids, equations, bar charts to probability. The reader can read the curriculum map to see how each piece stands and its relation to other pieces. The kind of teaching and learning for each piece is unique and needs planning way ahead. Third, this book presents different *situations* for the attempt—Math-Problem-Posing-in-action. The situation could be individual remedial teaching, small group, or a whole class; student-teaching or district-based teacher training; it could happen in and out of school or at home. Finally, problem posing assumed different faces depending on participants, math content, and the situation's intention of which the investigators took action. Results were reported and discussed for readers to imagine how that can happen in future investigations using more variables.

Although there is data analysis on tasks in varying contents than the chosen piece(s) reported in each chapter, the author include only selected topics for learning/teaching/parenting in one chapter that will not appear in the other chapters. After this selection, the math topics presented do not overlap, and the analyses of other examples will be reported elsewhere. A curriculum map is drawn to show the overall structure (see Fig. 3). In Fig. 3, the boxes contain math content pieces under four strands: D for data handling and probability, A for algebra, N for numbers and measures (N1, N2), where N1 is subdivided into numbers and computation, and N2 is also subdivided into different measures. Finally, S stands for shapes and space. Under these branches, the boxes contain the selected pieces of math content, from chapter two to chapter fourteen (C2–C14). The last number inside the box stands for grade (e.g. under D, probability, Grade 11). Blue, green, and orange boxes represent sections I, II, and III, respectively.

The book attempts to illustrate learning/teaching/parenting via math problem posing is feasible. The cases showed possibilities and difficulties in trying MPP for different math topics in different strands, at varying grade levels, and/or for participants from different backgrounds under various conditions of partnerships with a conscientious teacher educator.

This chapter is an overview chapter and three sections follow. They are students' learning (C2-C6), teachers' professional growth (C7-C10), and parental involvement (C11-C14).

In Part I, this MTE will be named "Susan" when she worked with a teacher and studied how students learn MPP. In Part II, she will be named "Professor Leung"

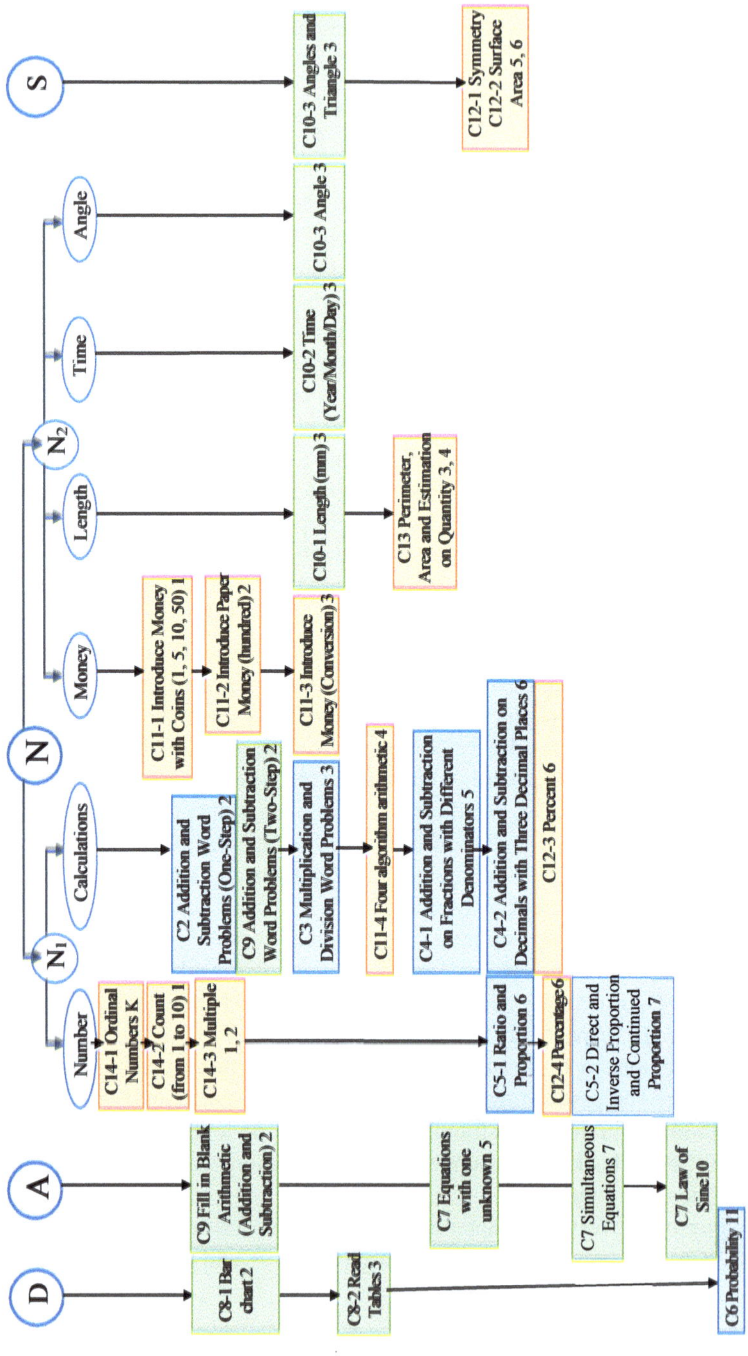

Fig. 3 Curriculum map for chapter two to chapter fourteen (C2–C14)

in teachers' training sessions or when she teaches prospective teachers. She will be named "Ms. Leung" when she borrowed classes and taught MPP for teachers (in-service, pre-service) to observe. Therefore, she assumed the roles of Professor Leung or Ms. Leung when doing teacher training. In Part III, this MTE will be named "Mytle" (a name spelled with letters M, T, and E), when she worked with her parents. She also took the role of a parent when she conducted family math camps.

Over 30 years of teaching at a university, she began as a professor (inexperienced in elementary school teaching) who found herself inadequate when she taught research ideas on how to teach children in elementary schools. She then created chances to teach (K-12) and stepped in and out of the Ivory Tower, assuming the roles of a researcher or a practitioner. In reflection, she referred to Shulman (1986) to remind herself to practice and "understand" to teach others how to teach. She also remembered five metaphors (Silver & Lunsford, 2017) and recalled her actions in using metaphors to bridge research and practice. In *osmosis* and *border-crossing* she and teacher/parents learn from each other. After her actions, she witnessed learning of herself and a group of students, teachers, and parents. The co-learners walked together through a time tunnel, to complete a wonderful and fruitful journey.

It was such an unforgettable journey.

References

Boston, M. D., & Smith, M. S. (2009). Transforming secondary mathematics teaching: Increasing the cognitive demand of instructional tasks used in teachers' classrooms. *Journal for Research in Mathematics Education, 40*, 119–156.

Brown, S. I., & Walter, M. I. (1983). *The art of problem posing*. Franklin Institute Press.

Butts, T. (1980). Posing problems properly. In S. Krulik & R. Reys (Eds.), *Problem solving in school mathematics* (pp. 23–33). National Council of Teachers of Mathematics.

Cai, J. (Ed.) 2017). *Compendium for research in mathematics education*. National Council of Teachers of Mathematics.

Christou, C., Mousoulides, N., Pittalis, M., Pitta-Pantazi, D., & Sriraman, B. (2005). An empirical taxonomy of problem posing processes. *ZDM, 37*(3), 149–158.

Ellerton, N. F., & Ellerton, H.D. (1987). Mathematics and chemistry problems created by students. In *Proceedings of the second international seminar: Misconceptions and educational strategies in science and mathematics* (Vol. 3, pp. 131–136) Cornell: Cornell University.

English, L. (2003). Mathematical modeling with young learners. In *Mathematical modelling* (pp. 3–17). Woodhead Publishing.

Freudenthal, H. (1973). *Mathematics as an educational task*. D. Reidel.

Goldin, G. A., & McClintock, C. E. (1979). *Task variables in mathematical problem solving*. Franklin Institute Press.

Grouws, D. A. (1992). *Handbook of research on mathematics teaching and learning*. Macmillian.

Henningsen, M., & Stein, M. K. (1997). Mathematical tasks and student cognition: Classroom-based factors that support and inhibit high-level mathematical thinking and reasoning. *Journal for Research in Mathematics Education, 28*(5), 524–549.

Hsu, H. Y., & Silver, E. A. (2014). Cognitive complexity of mathematics instructional tasks in a Taiwanese classroom: An examination of task sources. *Journal for Research in Mathematics Education, 45*(4), 460–496.

Ingvarson, L., & Rowley, G. (2017). Quality assurance in teacher education and outcomes: A study of 17 countries. *Educational Researcher, 46*(4), 177–193.

Kilpatrick, J. (1978). Variables and methodologies in research on problem solving. In L. L. Hatfield (Ed.), *Mathematical problem solving* (pp. 7–20). ERIC/SMEAC.

Krutetskii, V. A. (1976). In J. Kilpatrick & Wirszup (Eds.), *The psychology of mathematical abilities in schoolchildren*. Translated from Russian by J. Teller (Trans.). The University of Chicago Press.

Lester, F. K. (Ed.). (2007). *Second handbook of research on mathematics teaching and learning: A project of the National Council of Teachers of Mathematics*. IAP.

Leung, S. S. (1991). Building connections to the Pythagorean Theorem: An example of teachers' treatment of textbook problems. In M. K. Heid & G. W. Blume (Eds.), *Mathematics Education: Making connections* (pp. 41–47). Pennsylvania Council for the Teachers of Mathematics [Handbook of PCTM]

Leung, S. S. (1993). Mathematical problem posing: The influence of task formats, mathematics knowledge, and creative thinking. In I. Hirabayashi, N. Nohda, K. Shigematsu, & F. Lin (Eds.). *Proceedings of the 17th international conference of the International Group for the psychology of Mathematics Education, Vol. III* (pp. 33–40). Author.

Leung, S. S. (1994). On analyzing problem-posing processes: A study of prospective elementary teachers differing in mathematics knowledge. In J. P. da Ponte & J. F. Matos (Eds.), *Proceedings of the 18th international conference of the international group for the psychology of mathematics education* (Vol. III, pp. 168–175). Author

Leung, S. S. (1996). Problem posing as assessment: Reflections and Re-constructions. *The Mathematics Educator*, *1*(2), 159–171, Singapore.

Leung, S. S. (1997). On the role of creative thinking in problem posing. *Zentralbatt für Didaktik der Mathematik*, *97*(3), 81–85 (International Reviews on Mathematical Education), Germany.

Leung, S. S. (2009). Research efforts on probing students' conceptions in mathematics and reality: Structuring problems, solving problems, and justifying solutions. In L. Verschaffel, B. Greer, W. Van Dooren, & S. Mukhopadhyay (Eds.), *Words and worlds: Modelling verbal descriptions of situations* (pp. 213–225). Sense Publishers.

Leung, S. S. (2013). Teachers implementing mathematical problem posing in the classroom: Challenges and strategies. *Educational Studies in Mathematics, 83*(1), 103–116.

Leung, S. S. (2017). Teaching problem solving in Taiwan for graduate students in institute of education. In M. Stein (Ed.), *A Life's time for mathematics education and problem solving* (pp. 261–275). WTM Verlag für wissenschaftliche Texte und Medien Münster.

Leung S. S., & Silver, E. A. (1997). The role of task format, mathematics knowledge, and creative thinking on the arithmetic problem posing of prospective elementary school teachers. *Mathematics Education Research Journal, 9*(1), 5–24, Australia.

Leung, S. S., & Wu, R. (1999) Posing problems: Two classroom examples. Mathematics teaching in the middle grades. *National Council of Teachers of Mathematics, 5*(2), 112–117, USA.

Leung, S. S., & Wu, R. (2000). Sharing Problem Posing at Home Through Diary Writing: Case of Grade One. *The Australian Mathematics Primary Classroom, 5*(1), 28–32.

Mestre, J. P. (2002). Probing adults' conceptual understanding and transfer of learning via problem posing. *Journal of Applied Developmental Psychology, 23*(1), 9–50.

Ministry of Education. (1993). *Curriculum standard for elementary school*. Taijie (in Chinese).

Ministry of Education. (2014). *Curriculum guidelines of 12-Year basic education general guidelines* (in Chinese)

Ministry of Education of Taiwan. (2008). *General guidelines of grades 1–9 curriculum for elementary and junior high school education* (in Chinese).

Ministry of Education of Taiwan. (2018). *Curriculum guidelines for the 12-year basic education elementary school, junior high school, and upper secondary school the domain of mathematics* (in Chinese).

Ministry of Education of Taiwan. (2020, February). *Math curriculum guidebook -mathematics* (in Chinese).

National Council of Teachers of Mathematics (NCTM). (1989). *Curriculum and evaluation standards for school mathematics*. National Council of Teachers of Mathematics.

National Council of Teachers of Mathematics. (1991). *Professional standards for teaching mathematics*. Author.

National Council of Teachers of Mathematics. (2000). *Principles and standards for school mathematics*. Author.

Niss, M., & Højgaard, T. (2019). Mathematical competencies revisited. *Educational Studies in Mathematics, 102*, 9–28.

Polya, G. (1945). *How to solve it; a new aspect of mathematical method*. Princeton University Press.

Polya, G. (1954). *Mathematics and Plausible Reasoning (Volume II): Patterns of Plausible Inference*. Princeton University Press.

Reitman, E. (1965). Cognition and thought. Wiley. Richard Serra sculpture: Forty years. *June 3-September 10, 2007. Exhibition brochure*. MOMA.

Schoenfeld, A. (2024, March). Book review: Charting a career and its influence. J. Cai, G. J. Stylianides, & P. A. Kenney (Eds.), (2023). Research studies on learning and teaching of mathematics: Dedicated to Edward A. Silver. *Educational Studies in Mathematics*, online. https://doi.org/10.1007/s10649-024-10316-6

Shulman, L. (1986). Those Who Understand: Knowledge Growth in Teaching, *Educational Researcher, 15*(2), 4–14.

Silver, E. A. (1994). On mathematical problem posing. *For the Learning of Mathematics, 14*(1), 19–28.

Silver, E. A. (2013). Problem-posing research in mathematics education: Looking back, looking around, and looking ahead. *Educational Studies in Mathematics, 83*(1), 157–162.

Silver, E. A., & Lunsford, C. (2017). Linking research and practice in mathematics education: Perspectives and pathways. In Jinfa Cai (Ed.). *Compendium for research in mathematics education*, (pp. 28–47). National Council of Teachers of Mathematics.

Simon, H. A. (1973). The structure of ill-structured problems. *Artificial Intelligence, 4*(3–4), 181–201.

Stein, Smith, Henningsen & Silver, E. A. (2010). *Implementing standards-based mathematics instruction: A casebook for professional development* (2nd ed.).

Stoyanova, E., & Ellerton, N. F. (1996). A framework for research into students' problem posing in school mathematics. *Technology in Mathematics Education, 4*(7), 518–525.

van den Heuvel-Panhuizen, M., Middleton, J. A., & Streefland, L. (1995). Student-generated problems: Easy and difficult problems on percentage. *For the Learning of Mathematics, 15*(3), 21–27.

Yang, K. L., Hsu, H. Y., & Cheng, Y. H. (2022). Opportunities and challenges of mathematics learning in Taiwan: A critical review. *ZDM—the International Journal on Mathematics Education, 54*, 569–580.

Part I
Students' Learning

Chapter 2
A Case Study of Learning Addition and Subtraction Word Problems Through Problem Posing in Remedial Instruction

1 Introduction

Other than problem-solving, problem-posing is also an important mathematical activity (Silver, 1994) and as well a great component of the mathematics curriculum (MOE, 2018; NCTM, 1989). In Taiwan, examples of mathematical problem posing were given in curriculum guidebooks (NAER, 2020). Despite its significance in research and practice how children use math problem posing to learn math has not been thoroughly studied. This study, together with all other studies given in this book, was carried out in Taiwan. Taiwan outperformed many regions in mathematics performance in large-scale studies like PISA and TIMSS (Mullis et al., 2019; OECD, 2023). However, results also indicated that there is a huge variability in performance within this region. The variability is also great, even within a class. Students' math ability ranges from very high to very low. Hence, it is important to attend to those low achievers (*subject variable*, Kilpatrick, 1978) before it is too late. In this chapter, the investigators started from the elementary grade levels. And among the many learning problems in elementary mathematics, understanding word problems is a tough one.

Here is what happened...

A math teacher educator (Susan) and an elementary school teacher (Glory) worked together. The target here is addition and subtraction word problems (*task variable*, Kilpatrick, 1978). This topic is especially important in children's ability to pose addition and subtraction problems when doing problem solving involving numbers. There is substantial literature on how children solve problems but less on how they pose problems (English, 1998). Children's understanding of situations matching to operate symbols such as "+" or "−" was limited. As they solve textbook problems, the problems are already divided into sections on one single type for practice. After some practice on a single type of problem, they move to practice another type in the next chapter. In real life, problems appear in mixed categories, and they are meant for understanding before one can try problem solving. For example, among addition and subtraction word problems, the semantic structures may be CHANGE

S. S. Leung, *Mathematics Problem Posing in Action*,
https://doi.org/10.1007/978-981-96-8480-9_2

or COMBINE problems. When children do not have a clear one-to-one mapping between the problem situation and the operation, then understanding is even more difficult. This is also true for other semantic types such as COMPARE problems.

In this study, the problem-posing use is for remedial teaching (*situation variable*, Kilpatrick, 1978). The kind of problem posing is specifically restricted to the posing of addition and subtraction word problems, to match to one of two given line segment diagrams for CHANGE, COMBINE, or COMPARE word problems. The purpose of this study was to understand how a child learned addition and subtraction word problems by carrying out this kind of problem posing in real classroom settings in remedial teaching. The research question is: What were the changes (if any) in a grade two child after remedial teaching, in problem solving performance and attitude?

2 Background Literature for Actions

A knowledge of children's number learning is needed, for school teachers to understand how students learn addition and subtraction. However, the math teacher educator (in this case, Susan) believed that the background needed for the teacher (Glory) to assist students should be directly related so that the literature would not confuse the teacher. This section included what Susan shared with Glory, which included the background literature for taking action. In the search for literature on addition and subtraction word problems, Susan and Glory met for reading on the learning of whole number addition and subtraction, and, problem posing with line segment diagrams for addition and subtraction situations.

2.1 On the Learning of Whole Number Addition and Subtraction

Nesher et al. (1982) presented empirical data for different categories of addition and subtraction word problems; with different performances for children. They used results to suggest the children's developmental levels of word problem-solving ability related to empirical and logical knowledge structure. Finally, they used a framework to demonstrate how these levels account for children's different performances in different categories. Word problems on addition and subtraction are given in three semantic types (CHANGE, COMBINE, COMPARE), adding to 14 categories. Within each semantic type, the "unknown-known" is different, thus, resulting in a different number of problems.

CHANGE

– Increase:

1. Unknown *given*; 2. Unknown *change*; 3. Unknown *result.*

– Decrease:

4. Unknown *given*; 5. Unknown *change*; 6. Unknown *result.*

COMBINE

7. Unknown *part*; 8. Unknown *result.*

COMPARE

– More:
9. Unknown *referent*; 10. Unknown *compare*; 11. Unknown *difference.*

– Less:
12. Unknown *referent*; 13. Unknown *compare*; 14. Unknown *difference.*

A computation, "24–18 = 6", may represent different situations.

CHANGE: 6. Unknown *result*
There are 24 pencils and I used 18, how many were left?
COMBINE: 7 Unknown *part*
A basket holds 24 red and green apples. 18 of them are red. How many are green?
COMPARE-More: 11 Unknown *difference*
There are 24 boys and 18 girls in the class. How many more boys than girls in this class?

The computation "24–18 = 6" for the above three word problems is the same. However, the understanding of the semantic structure of the above three-word problems is not at the same difficulty level. Thus, children may have some difficulty in solving the word problem, even though they can compute "24–18 = 6".

In prior studies, the investigators used children's percentage of correct in the above 14 categories to suggest the varying amount of difficulty levels in understanding these word problems (e.g., Nesher et al., 1982). For example, CHANGE was found to be easier than COMBINE, and COMBINE was found to be easier than COMPARE. When addition and subtraction word problems are not just a computation and are given in real-world situations children make sense of them gradually and undergo different stages. According to Fuson (1992), children's developmental progression was in three conceptual structures: Single representation of an added or the sum, abbreviation of sequential counting, and derived facts. Due to varying math abilities and understanding of the situations, children solved addition and subtraction word problems in many different ways. The different ways children solved them, with or without teaching aids, were also found in interviews from Cognitive Guided Instructions. They are direct modeling, counting, and derived facts (Carpenter et al., 1996). In all, results on how children learned this topic have implications for teaching (e.g. see Fuson, 1992).

2.2 *Problem Posing with Line Segment Diagrams for Addition and Subtraction Situations*

Susan shared with Glory a study on problem posing (see English, 1998). In this study, the investigators developed a problem-posing program (with sixteen 45-min sessions) and had grade 3 children pose problems in formal and informal contexts. Formal contexts meant giving addition and subtraction number sentences and children were to pose word problems. Informal contexts meant non-goal-specific situations that were stories without problems, stimulus pictures, and literature. Though children improved in generating problems over the 16 sessions, she found that children had difficulties posing a wide variety of problem situations matching addition and subtraction number sentences in formal contexts. For the most part, children posed CHANGE/part-part-whole in the program activities, by altering the contexts of their problems. English (1998) commented that the activities of the program were insufficient to broaden the children's interpretations of these formal statements. She suggested exploring more so that children can extend basic CHANGE/part-part-whole types. Therefore, materials with problems that extend to a variety of semantic types, worksheets, or teachers' prompts must be carefully designed to suggest children pose such problems.

When teachers read literature on how children solve addition and subtraction word problems, it also helps them how to teach. For representing addition and subtraction word problems, teachers and children may use teaching aids like unifix cubes to model the situations. These models were found to be useful in interviews (see CGI, Carpenter et al., 1996). In solving, children can manipulate objects, draw, count out loud, or, write numbers and symbols. For paper and pencil tasks, line segment diagrams can be used to show the relationships of any two quantities. Two typical line segment diagrams representing addition and subtraction are given below.

As above, the left line segment diagram can be used to represent a CHANGE situation (e.g., buy 18 chocolates then 6 more resulting in 24 in total) or a COMBINE situation (e.g., a bag of 24 apples with 18 red and 6 green). The right segment diagram can be used to represent a COMPARE situation (e.g., A pencil is 22 cm long and a pen is 16 cm. The pen is shorter than the pencil by 6 cm).

In this study, the aim is to address target addition/subtraction word problems that children often miss, the teacher had to tailor a one-to-one remedial teaching program using problem-posing and problem-solving. Along with presenting a word problem in context and making it interesting, Glory and Susan decided to use picture books, which were especially useful for children in grade one and grade two.

3 Method

The research was conducted for one semester, to complete a second-grade remedial teaching unit with posing problem. The researchers met regularly to attend to matters arising from this individual teaching. They designed a sequence of tasks for problem-posing and problem-solving.

3.1 Participants

The teacher: Ms. Glory

The teacher is also one of the researchers of this case study. Glory was the class teacher of the class who also taught math. Glory taught in this primary school for fourteen years after obtaining her degree in math and science education. She actively participated in professional development projects. In one professional development program, she met Susan (a math teacher educator) and shared her teaching problems. They made a plan for a research study. Glory also completed a part-time postgraduate course in case study and action research and finally extended the work into a graduate thesis. She assumed the roles of the instruction designer, teacher, and observer in this remedial teaching program. The whole process of the remedial teaching program was audio-recorded (and videotaped) for more objective data analysis. Glory had discussions with her thesis advisor and two colleagues per fortnight. Additional meetings were scheduled when needed. The discussions helped to minimize analysis by subjective thinking.

Glory believed that learning attitude is a persistent and consistent behavioral tendency that an individual maintains toward learning, and it will be affected by the relationship between the individual's cognition, emotion, and action. Learning attitude is very broad, which may include methods, plans, habits, desires, processes, and motivation, and can refer to attitude in class or exams. She considered attitude a psychological process, a potential variable, and difficult to be directly measured by written tasks. For a child who was only in second grade, she tried to evaluate through observations to explicit behaviors. Glory decided to use long-term observations during remedial teaching to capture the child's learning attitude instead of a written test.

The target student: Theo

In her class of 27 children, there were 7 who scored at the bottom 27% among all in mathematics. Two were chosen as participants for remedial teaching. Both students failed to solve word problems but had no problem in computations or in verbally expressing themselves. On top of that, their parents agreed to participate during the teacher's suggested time slots. The remedial teaching was done individually by the same teacher and thus, using two different time slots. In this chapter, one target student will be reported (the other student shared similar difficulties and was reported elsewhere in a longer report).

Theo was an extroverted boy with a happy family. However, he was easily distracted and could not focus on math problems. He had difficulty understanding word problems and thus lacked learning confidence. Theo did not have sufficient independent learning skills. He would immediately turn to the teacher or copies answers from other students' when he found the math problems difficult. Theo's mum helped him with math learning and sent him to tutoring programs, but Theo was still struggling with the subject, especially word problems.

3.2 Design of Remedial Teaching Program

First, Ms. Glory used a picture book as a context to present addition and subtraction word problems.

Children loved stories. A picture book is especially suitable for preschool to second-grade children. The content includes pictures, words, and themes. Drawings are used as the main form of expression; the text features are easy to understand, which aligns with children's understanding; usually, the theme focuses on children's concerns and interests. Picture books are beneficial to arouse readers' interest, and contextual understanding, and promote the learning experience. There are many picture books but few that Ms. Glory would think suitable for learning mathematical concepts; but, more suitable for life experience or culture. Glory decided to use a story that is appropriate for children's life experiences or culture. The teacher chose one book called "*The Happy Dragon Prince*".

> I believe that children love stories so I decided to use picture books to attract them. When I use the projector to play the stories there are sound effects to go with motion pictures. This will make the story more interesting. Then I can try math problem posing.

Addition/subtraction problems in primary math curriculum.

Second, she read literature and studied the types of addition/subtraction word problems (Moser & Carpenter, 1984). She then presented problems in the Dragon Prince context.

> I read Nesher, Greeno, and Riley (1982) from a graduate seminar. Later, I found similar descriptions of types and classifications from the teacher guide given in the textbook. This will give me a good start.

She listed out three semantic types in her notebook: CHANGE, COMBINE, and COMPARE with a total of 14 sub-categories.

CHANGE

- Increase: 1 Unknown given; 2 Unknown change; 3 Unknown result
- Decrease: 4 Unknown given 5 Unknown change 6 Unknown result

COMBINE

7 Unknown part; 8 Unknown result

COMPARE

– More: 9 Unknown referent; 10 Unknown compare; 11 Unknown difference
– Less: 12 Unknown referents; 13 Unknown compare; 14 Unknown difference

However, she only chose 9 as the target problems (1, 4, 7, 9–14; underlined) for remedial teaching because children scored 100% in the remaining types.

Third, she designed a remedial teaching program.

The investigator developed a program for her remedial teaching and used a flowchart to explain it (see Fig. 1).

The remedial instruction included problem-solving and problem posing. The cycle was repeated nine times as there were nine types. In presenting the picture book, Glory played the video of *The Happy Dragon Prince* (Step 1).

In problem-solving, she provided students with a problem-solving worksheet of one type of addition/subtraction question (from the nine types) using the story as

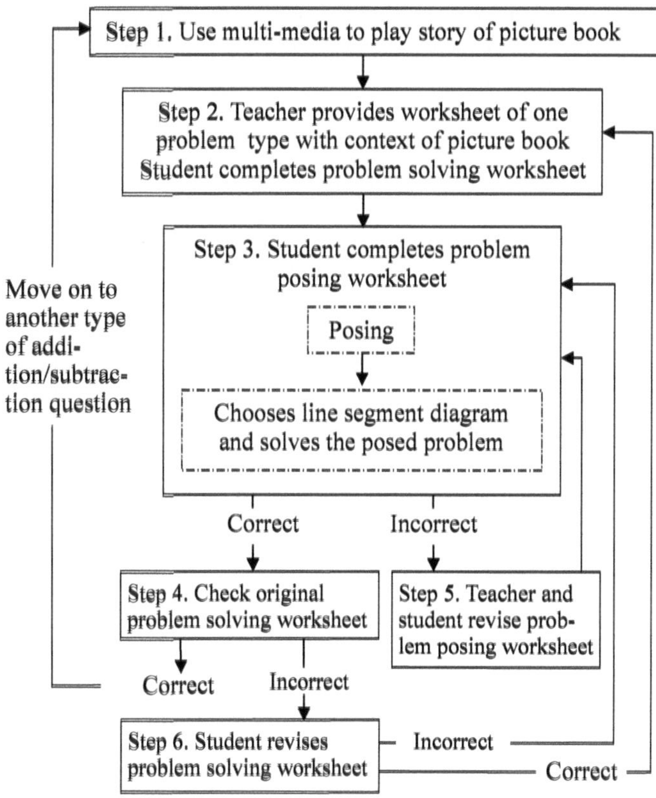

Fig. 1 Flowchart of Ms. Glory's remedial teaching program

a context. Students then independently completed the worksheet and checked their answers (Step 2).

In problem posing, children were asked to complete a problem-posing worksheet for the same type of question. They needed to pose a problem using a line segment diagram, and then solve that problem (Step 3).

Worksheets for problem solving and problem posing were given in the Appendix. Note that students did not need to draw any line segment diagram. Ms. Glory provided two for students to pick one from two each time they posed a word problem. Ms. Glory also provided concrete materials together with the line segment diagrams. Children used one cm cubes as a unit and the line diagram was drawn to scale (e.g. put 6 cubes next to a 6 cm line segment diagram). If the students could not complete the problem posing worksheet, Glory would correct the worksheet with the student and ask the student to redo problem posing (Step 3).

In evaluation, if the student could complete the problem posing worksheet, then they solved the problem (Step 4). If the solving was correct, they moved on to another type of addition/subtraction question (Step 2 again). If solving was incorrect then the student revised problem-solving worksheets (Step 6) until correct and moved on to another type of addition/subtraction questions (Step 2 again). However, when in evaluation, if posing was incorrect, the teacher and student revised (step 5) and redid problem posing (step 3 again).

3.3 *Data Collection and Analysis*

Instrument: Problem-solving test. Glory's nine target problem types and examples (see Table 1).

By referring to the above problem types, she also designed problem-solving and problem-posing worksheets. There is one example problem type in the problem solving worksheet (Appendix). It is an example problem type 13, COMPARE-Less: Unknown compare (see Table 1).

Data analysis: she compared the two test scores (pre-test and post-test).

During the interview, she used the four stages of problem-solving (Polya, 1945): Understand, Plan, Carry Out, and Look Back:

1. Circle the goal and underline statement (Understand),
2. Representation in figure (Plan),
3. Write the expression and solve (Carry Out).
4. Review and talk (Look Back), done by interview.

Data included: addition and subtraction word problems pre-test and post-test, problem-solving worksheets, problem-posing worksheets, learning diaries, researcher's teaching introspection record, and, audio-taping. The interviews were audio-typed and results were used to analyse changes qualitatively.

Table 1 Nine target problem types in tests and examples

Problem type	Unknown	Pre-test	Post-test
CHANGE-Increase	1 Unknown given, known change, known result	Sam had some bananas. Emma gave him 10 more. Sam now has 28 bananas. How many did Sam have?	A class had some students. 12 more students went to the class. It now has 25 students. How many students were at the beginning?
CHANGE-Decrease	4 Unknown given, known change, known result	Sam had some bananas. His friends ate 42 of them. Now he has 5 left. How many bananas did Sam have?	A class had some students. 21 students left the class. It now has 4 students. How many students were at the start?
COMBINE	7 Unknown part, known result	Sam and Emma have 28 bananas in total. Emma has 16 bananas. How many does Sam have?	A class has 24 students. 18 of them are boys. How many girls are in this class?
COMPARE-More	9 Unknown referent, known compare, known difference	Emma has 54 bananas. She has 18 more bananas than Sam. How many bananas does Sam have?	There are 14 boys in the class. The number of boys is 6 more than girls. How many girls are in this class?
	10 Known referent, Unknown compare, known difference	Sam has 17 bananas. Emma has 4 more bananas than Sam. How many banana does Emma have?	There are 5 boys in the class. The number of girls is 8 more than boys. How many girls are in this class?
COMPARE-Less	11 Known referent, known compare, unknown difference	Sam has 37 bananas. Emma has 24 bananas. How many more bananas does Sam have compared to Emma?	There are 17 boys and 6 girls in the class. How many more boys than girls in this class?
	12 Unknown referent, known compare, known difference	Emma has 25 bananas. She has 7 less than Sam. How many bananas does Sam have?	There are 10 boys in the class. The number of boys is 2 less than girls. How many girls are in this class?

(continued)

Table 1 (continued)

Problem type	Unknown	Pre-test	Post-test
	13 Known referent, unknown compare, known difference	Sam has 48 bananas. Emma has 15 less bananas than Sam. How many does Emma have?	There are 16 boys in the class. The number of girls is 8 less than boys. How many girls are in this class?
	14 Known referent, known compare, unknown difference	Sam has 23 bananas. Emma has 18 bananas. How many fewer bananas does Emma have compared to Sam?	There are 17 girls and 8 boys in the class. How many fewer boys than girls are in this class?

4 Results and Discussions

4.1 Change in Performance and Problem-Solving Process

Theo changed after MPP teaching. There are three parts: (a) changes in pre-test and post-test, (b) changes from one-to-one remedial instructions and (c) changes in problem-solving processes.

(a) Findings on addition and subtraction word problems test scores (Comparison of Pre-test and Post-test results)

Theo answered 13 questions correctly in the pre-test and 41 in the post-test. It was shown that Theo found the right answers to 28 more questions after the remedial teaching program. Theo could also self-correct all of his mistakes in the post-test interview. In all, he improved after MPP teaching.

(b) Findings from a one-to-one interview (remedial instructions)

Results included examples from different steps in solving and posing plus revisions. There are three parts: (i) Theo had difficulty distinguishing target questions from known conditions, (ii) mixed up larger and smaller numbers, and, (iii) Theo's choice of diagram influenced by previous types.

(i) Theo had difficulty distinguishing target questions from known conditions (Step 2: Problem solving).

When Theo attempted his problem-solving worksheet for the first time, he did not complete questions in the following three sections: Find the target question, Re-state known conditions, and, Complete line segment diagrams. To help Theo understand the problems, the teacher used two pencils of different lengths to represent two-line segments. In his learning diary of the day, Theo wrote that he found the problems very hard. "*I do not think I can solve it. I cannot understand the problems.*"

(ii) Theo often mixed up larger and smaller numbers (Step 3: Problem posing).

When Theo posed problems and drew line segment diagrams, he tended to imitate the structure of worked examples, while ignoring the size of numbers and their relation to the real-life scenario in the problems he posed.

The problem posed by Theo: Lily has 10 books. Tom has 15 books. How many books does Lily have more than Tom?

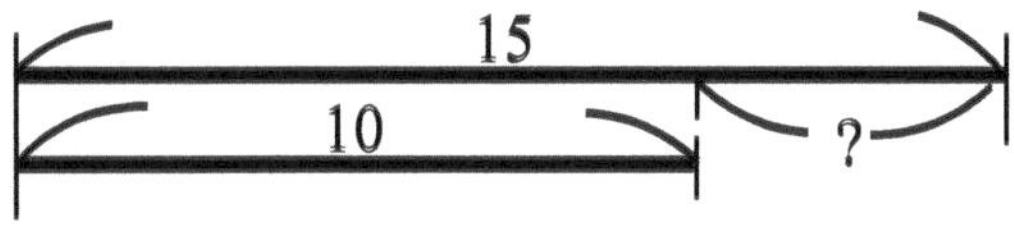

Theo did not realize that Lily had fewer books than Tom. In the line segment diagram, Theo forgot that the size of numbers should match the length of line segments.

The other example was: Mom brought 10 books. She brought 15 more books than Dad. How many books did Dad bring?".

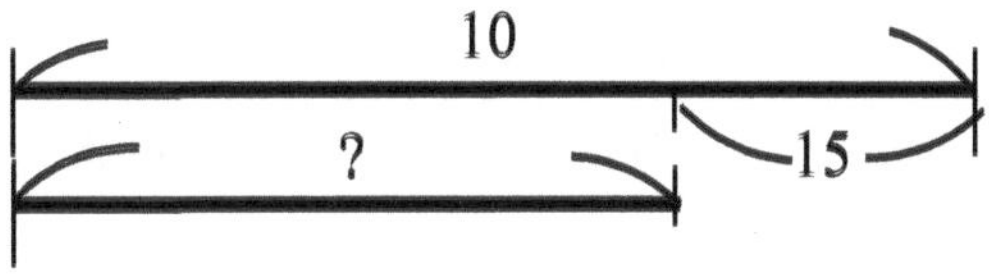

When he was trying to solve his problem, Theo was confused by the keyword "more" in the problem. When posing concerning line segment diagrams, Theo often mixed up with size of numbers and their relative length in the line figure.

(iii) Choice of diagram influenced by previous types (Step 5. Teacher and student revise problem posing worksheet).

When Theo chose line segment diagrams, he did not think independently, and he was likely to be influenced by previous types of problems. Theo's answers on the problem-posing worksheet and conversations during the teaching process of a combination type of line segment diagram are presented below.

Problem: Doraemon has 10 friends. 5 of his friends like reading books. How many friends do not like reading books?

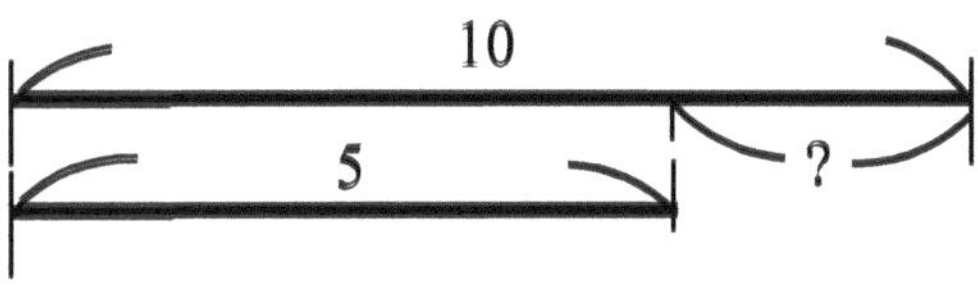

Teacher: Does your question want to compare the number of all friends with friends who don't like reading books?

Theo: No!
Teacher: Your problem divides friends into…
Theo: Who reads books and who doesn't?
Teacher: This side presents Doraemon's friends who like reading books (a pencil in the left hand) and this side presents his friends who don't like reading (a pencil in the right hand). Putting them together (combining two pencils into one line) …
Theo: All friends!
Teacher: Correct! Think of friends on this side who like reading (shaking the pencil in the left hand) as a line segment, and think of friends on the other side who don't like reading (shaking the pencil in the right hand) also as a line segment, then all friends are (combining two pencils as one line) …
Theo: Putting two line segments together.

Theo was influenced by previous problem types. He tended to use the largest number in the question as his longest line segment, and as a result, he filled the difference and compared values into the wrong positions in the line segment diagram again.

(c) change in problem-solving processes

There are altogether three parts: (i) UNDERSTAND, (ii) PLAN, and, (iii) CARRY OUT.

(i) UNDERSTAND STAGE: Inert (before remedial instruction) but made an effort to understand well (after remedial instruction).

Ms. Glory used multiple-choice items to see if he understood (restate givens, search for goals, and identify relational statements). There were three options to choose for each item (1, 2, and 3).

First, how Theo restated givens in the problem. (Type 14)

Theo could restate conditions given in the problem and options to answer, did not just merely refer to the problem description word by word. Below was a comparison between pre-test and post-test.

Type 13. COMPARE (Less) Unknown Compare

Pre-test
The length of the rope is 48 cm. The ruler is 15 cm shorter than the rope. How long is the ruler?

Which of the following did the problem tell you?

1. Rope is 15 cm shorter than the ruler
2. Ruler is 15 cm long
3. Rope is 48 cm long

Post-test
There are 26 monkeys in the zoo, and there are 13 fewer peacocks than monkeys.

How many peacocks are there in the zoo?

Which of the following did the problem tell you?

1. there are 26 monkeys in the zoo
2. there are 13 peacocks in the zoo
3. the number of monkeys is 13 less than that of peacocks

From the above figures we can see that in the pre-test, Theo used matching of options (1, 2, 3) to match conditions given in the problem but did not understand that the piece of string was represented by an actual number and the question asked for an unknown number that represented the length of a ruler. In the post-test, the order of reading the question and of reading the answer options was different. Theo checked the meaning of the question first and then, considered the three options and selected option 1. Finally, he explained why did not select the other two options.

To Theo, the way of answering questions in the pre-test is to check the text content of the questions, not to understand the semantics. For example, he said: "the length of the rope is 48 cm", which was inconsistent with option 2 (the length of the ruler is 15 cm), so option 2 was not selected. It shows that he does not understand the given conditions in the question refers to an affirmative narrative with definite numbers.

In the post-test, Theo understood the meaning of the questions and options, for example: "Option 2 is not selected, because the peacock in the question is unknown, and it is the question to be asked in the question." Theo already knew that the goal was "the number of peacocks", and the option "the number of peacocks is known" was wrong. "option 3 means that there are fewer monkeys, which is exactly the opposite of the meaning "There are 13 fewer peacocks than monkeys", so I don't choose option 3."Theo knew that the option meant that there were few monkeys, which was inconsistent with the topic, so he didn't choose it.

Second, how Theo searched for the goal in the problem CHANGE-Increase Unknown given. (Type 1)

In the pre-test, Theo was asked to find the question or the goal in problem-solving. He read the three options one by one. After hesitating and considering the first, second, and third, Theo did not choose any one of them. When the investigator asked why Theo did not choose, Theo said none was appropriate. After remedial teaching and when Theo completed the post-test, Theo matched the option to the unknown given.

In the pre-test, Theo knew that the question asked "How many birds were there in the tree", but when answering the question, he read the question once and found that there was no similar description, so he did not choose it. For example, he did not choose the first option: It is "there were some birds on the tree" so didn't choose.

um… (he thought for a while and then shook his head to indicate that he didn't choose). It showed that Theo didn't know the target of the problem.

In the post-test, Theo understood the meaning of the question and the meaning of the options. For example, he said "Option 1: How many colored papers were there in the box" should be selected, because the title said how many colored papers were in the box, that's why he asked"; "option 2: How many more colored papers were out in the box?" How many-colored paper to put in", no, because we have been told that there are 20 sheets. Theo understands the content of the options and corresponds to the problem.

Type 1 CHANGE (Increase) Unknown Given.

Pre-test

Originally, there were some birds on the tree, and then 10 birds flew over, and finally, there were 28 birds on the tree. How many birds were there originally?

Which of the following did the problem ask you to answer?

1. How many birds originally?
2. How many more birds?
3. How many birds finally?

Post-test

There were some colored papers in the box, and 20 colored papers were put in, and finally, there were 47 colored papers in the box. How many colored papers were there in the box?

Which of the following did the problem ask you to answer?

1. How many colored papers were there in the box?
2. How many more colored papers were out in the box?
3. How many papers finally?

Third, how theo identified relational statements. (Type 14)

In the pre-test, Theo could not understand the question, nor the meaning of the question, so he checked the options word by word and found that he could not find a sentence with the same description, so he did not choose any of the three options.

For example: "It is "how long is afoot", not a foot. It is relatively long, so option 1 is not selected; it is "the ruler is 15 cm shorter than the rope", so option 2 is not selected; it is "the length of the rope is 48 cm", so option 3 is not selected."

In the post-test, Theo could identify the relational statement according to the meaning of the question. For example: "Because the question told us that there were 13 fewer peacocks than monkeys (point to the text), so the answer is option 3." Theo understood the meaning of the question and chose the correct option.

Type 13. COMPARE (Less) Unknown Compare

Pre-test

The length of the rope is 48 cm. The ruler is 15 cm shorter than the rope. How long is the ruler?

When comparing lengths, which of the following is correct?

1. Ruler is longer
2. Same
3. Ruler is shorter

Post-test

There are 26 monkeys in the zoo, and there are 13 fewer peacocks than monkeys. How many peacocks are there in the zoo?

When comparing the number of peacocks and the number of monkeys, which of the following is correct?

1. Same
2. More peacocks
3. More monkeys

(ii) PLAN STAGE: Incorrect figure representation affects the writing expression for solving (before remedial instruction); can link the meaning of problem, diagram, and set up expression (after remedial instruction). Theo did *figure representation* and then *set up expressions*. The two parts are given below.

The first part of Plan stage: how Theo did figure representation. (Type 14)

Type 13. COMPARE (Less) Unknown Compare

Pre-test

The length of the rope is 48 cm. The ruler is 15 cm shorter than the rope. How long is the ruler?

Which of the following represents the problem?

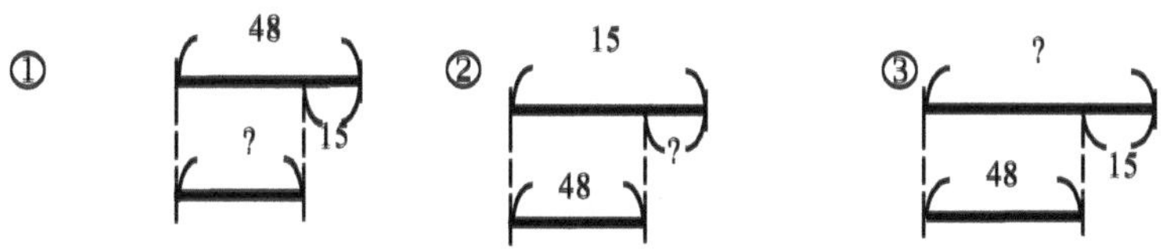

Post-test.

There are 26 monkeys in the zoo, and there are 13 fewer peacocks than monkeys. How many peacocks are there in the zoo?

Which of the following represents the problem?

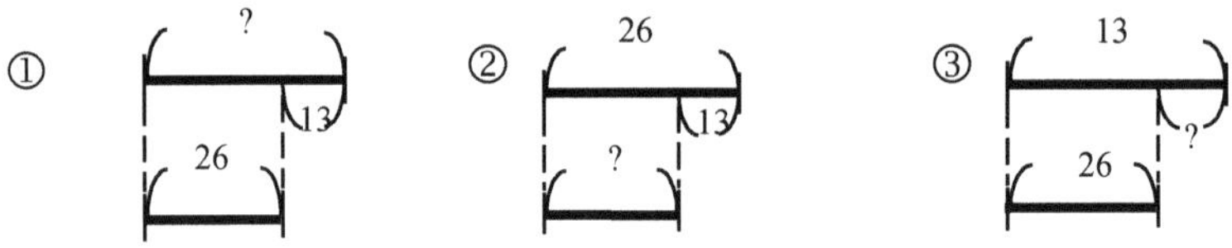

Second part of plan stage: how Theo sets up expressions. (Type 14)

In the Pre-test, Theo originally answered with the formula 48 + 15. Although he corrected the formula to 48–15 during the interview, he also chose another line diagram option number 3 which was completely inconsistent with 48–15. To Theo, the meaning of the problem, the line segment diagram, and the formula were unrelated to each other. Ms. Glory recorded Theo's thoughts during the pre-test.

However, during the post-test, Theo has been able to formulate according to the meaning of the question, and can also notice that the meaning of the formula and the line diagram must be consistent. For example, Theo saying option number 2 has the same meaning, "*26 is a monkey, and the question mark is a peacock. Ah, there are 13 peacocks less than monkeys here (point to 13 in option number 2), and the question mark is 26–13.*" He can combine the problem and line diagram link.

Type 13. COMPARE (Less) Unknown Compare

Pre-test

The length of the rope is 48 cm. The ruler is 15 cm shorter than the rope. How long is the ruler?

Please set up the expression and find the answer.

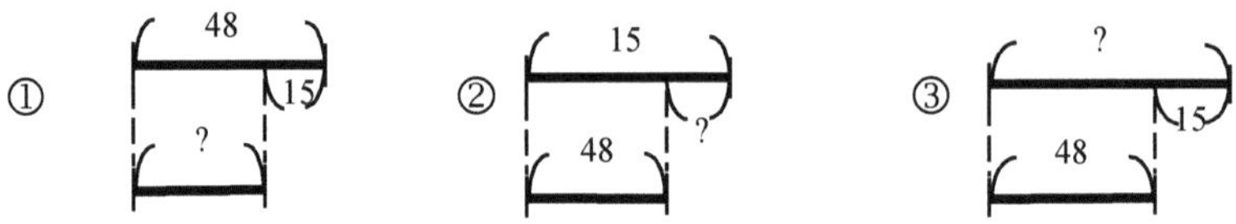

Post-test

There are 26 monkeys in the zoo, and there are 13 fewer peacocks than monkeys.

Please set up the expression and find the answer.

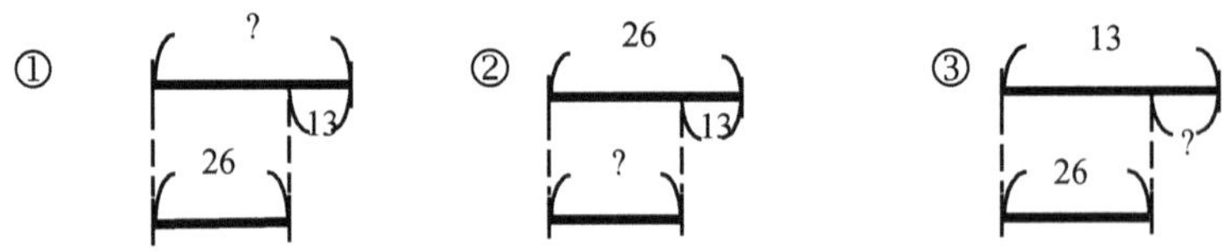

To sum up, before the remedial teaching, Theo had an incomplete understanding of the line segment diagram, only the size of the number will affect the length of the line segment, so the conversion between the text and the graphic representation cannot be correct; after the remedial teaching, Theo had a complete understanding of diagrams, and be able to form correct representation transformations between text, diagrams, and formulas.

(iii) CARRY OUT: Used partial representation to check if the answer was reasonable (Before remedial instruction) then used various representations to check (after remedial instruction).

Type 13. COMPARE (Less) Unknown Compare

Pre-test

The length of the rope is 48 cm. The ruler is 15 cm shorter than the rope. How long is the ruler? Please find the answer.

Post-test

There are 26 monkeys in the zoo, and there are 13 fewer peacocks than monkeys. How many peacocks were there? Please find the answer.

Combining Theo's pre-test and post-test, problem-solving performance, and interview content, it was found that there are two parts to the change in Theo's problem-solving ability.

First, in the comparison of the answers of the pre-test and post-test, after receiving remedial teaching every day, the number of questions answered correctly in both the test and the interview increased by 28 questions, and the 4 questions answered wrongly in the post-test were also correct during the post-test interview.

Second, in terms of the change from the interviews and problem-solving process, Theo was no longer indifferent in the stage of understanding the meaning of the question and drawing up the plan, and only used the strategy of checking the options and the content of the text question word-by-word to answer the questions. He changed the habit of not looking for line diagrams based on the meaning of the question, but only using the size of the number to intuitively determine the length and position of the line segment. In the stage of executing the plan, he will form the correct conversion between representations to solve the problem. In the stage of checking and reviewing, he would also use various interconvertible representations to verify the rationality of the answer.

4.2 *Theo's Change in Attitude*

Before MPP, he could not concentrate during the pre-test and missed 1/3 of the items. After the remedial instruction, Theo actively participated during the post-test and completed all items during the post-test. From one-to-one instruction, Ms. Glory witnessed his change, which was obvious. Originally, he would not concentrate on his work, read quickly, then gave his finger a big bite and shook his head.

During the pre-test interviews, Theo felt exhausted and the volume of his conversation was lower than usual. His problem-solving speed was slow. Ms. Glory questioned why he did not answer in the pre-test but answered correctly in the interview, he admitted that he did not pay enough attention. After the pre-test interview, Ms. Glory asked him to review the entire test paper to find out the easiest and most difficult questions. As a result, he just turned a few pages randomly without serious thinking to find the easiest and most difficult question. When reading six questions, he said that question 5 and question 6 were more difficult and his reason was normally the questions were from easy to difficult. He just did not want to pay attention to reading them again to review and compare them seriously. The experience made Theo feel insecure and frustrated about the math test. During the pre-test, Theo had negative learning attitude. After MPP instruction, he did not need any reminders and remembered what he was expected to do and even wrote notes to remind himself. He even turned through the previous pages to see if there were any missed items or pages. He was not careless anymore and became proud.

Please do not remind me, I know what to do next.

Theo: Miss, guess how many left? (Smiled with excitement).
Glory: Um ..... I guess four? (Blinked her eyes).
Theo: Only one! (Smiled with triumph. Continue to work immediately).

The change was also found during the interviews which were conducted after problem solving. Theo did not have much interest at the beginning and showed excitement in solving problems over time.

Ms. Glory exclaimed, "He was a worm and became an energetic dragon!"

Discussion on 4.1 and 4.2

About the change to Theo. From findings in test scores, and changes in posing and solving during individual interviews, Glory and Susan witnessed the improvement of Theo over time. In recall, they found that Theo (and the other child too) loved doing problem posing in this remedial teaching program after school and shared common responses during the individual teaching and interviews. Inclusion of problems in different semantic structures (Nesher et al, 1982) and sequencing them in the one-to-one interview was helpful in remedial teaching. The differences before and after MPP instructions were noted. Before MPP instruction, Theo could not figure out what the goals were; he mixed up large and small numbers and was much influenced by prior types of problems. After the instruction, he could spot out the goals and the givens. He did not just pick any option without thinking hard, liked what he did in the pre-test. Next, he selected the correct line segment diagram to match the situation given in the word problems and posed another problem. When he posed a problem using the line segment diagram he attended to his prior mistakes in the size of numbers. For example, in the case of problem type 11 (COMPARE-unknown difference), the "How many more" goal turned out to be problematic when he read the line segment for 24 bananas longer than the line representing 37 bananas. Thus, the line-segment diagram helped him to concretize the size of numbers.

Designing a teaching program. In this study, problems in needed semantic structures were included; and, were arranged from simple to complex–*according to Theo's prior individual performance.* For examples, the few problems that Theo's solved correctly were excluded. Only the nine categories that Theo found difficult were included in the one-to-one interview sessions. Problem posing is an inseparable part of problem solving (Leung, 2009). In this study, MPP was conducted together with problem solving (before and also after). Reading about how children solved problems and progressed from strategy one to another was most important. Note that the kind of problem posing was posing one that the student just solved and also matched a line segment graph. The investigators found that providing two line-segment diagrams as choices of representations for a word problem was helpful to a student like Theo. Ms. Glory believed that length was concrete to children who found word problems difficult to understand. Theo made improvements in using the diagrams to represent problems and to set up expressions to find the answers to the problems. In solving the posed problem, Theo identified the relational statements and then matched line segment diagrams. He was using the given information to write expressions and find solutions. Finally, his attitude changed also after he found able to understand the word problems.

5 Conclusion

A remedial program was designed to assist grade two children who did not understand word problems. A sequence of interviews were done by a school teacher on a one-to-one basis. One student was reported as a case who improved after using this program. This program included a variety of semantic types of addition and subtraction word problems. Two specific line segment diagrams were provided to assist posing problems and to represent addition and subtraction situations such as CHANGE, COMBINE and COMPARE. After continual effort in this remedial teaching program over time using solving, posing, and then solving, students improved in solving one-step problems with whole numbers up to two digits. In using problem posing as a means, there was a change in the child's (e.g. Theo) ability and attitude.

When children have difficulty in solving addition and subtraction word problems, what is required before knowing the operation required and the computation, is the identification of the problem structure given in the situations (Fuson, 1992). Teachers need to attend to children's progression or developmental stages (Nesher et al., 1982). In this study, the investigators addressed the results of a diagnosis of prior mistakes and attempted to use a carefully designed remedial teaching sequence of solving and posing programs (see Fig. 1 again). Posing problems was a means to meet this challenge.

The chapter closes with implications for future teaching and research. The teacher who learns children's way of solving is constructive to professional development (e.g., CGI, Carpenter et al., 1996). The teacher learned how to conduct teaching using control of attention span (e.g. 20 min here) to avoid fatigue or loss of enthusiasm. Children loved the use of picture book contexts for word problems. Kilpatrick (1978) mentioned *task, subject, and situation* as variables in problem solving. Here, the three variables were also considered, the understanding of *addition and subtraction word problems* for *low-achievers* (e.g. Theo) was made concrete under Glory's *MPP remedial program.* More attempts on other variables are reported in the following chapters.

Appendix: Worksheets (for Problem-Solving and Problem-Posing)

1. Problem solving worksheet

   Name: Date:

   > The mountain god can spray 58 flowers at a time, and the fire-breathing dragon prince can spray 12 fewer flowers at a time than the mountain god. How many flowers can the fire-breathing dragon prince spray at one time?

   1. Please circle the sentence that asks the question.
   2. Please find the sentences in the title that are useful to solve the problem, and underline the sentences.
   3. Please fill in the meaning of the question in the appropriate line diagram.

   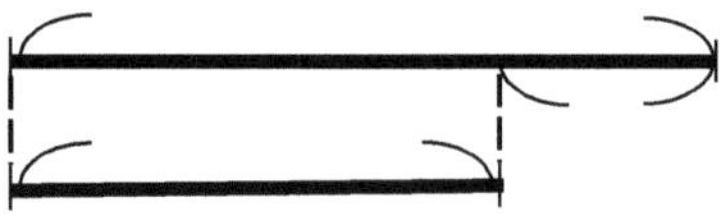

   4. Please write the expression and answer.

2. Problem posing worksheet.

   Name: Date:

   1. Please pose a problem; think of a topic with "…less than …".
   2. Please choose a line segment diagram to match the meaning of the topic in your problem.

   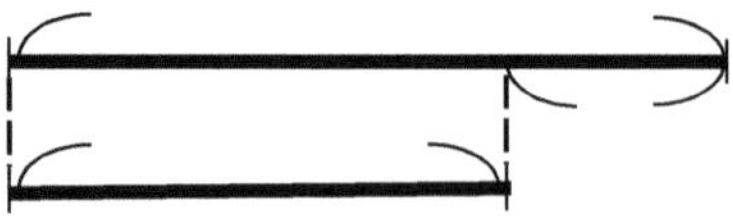

   3. Please write an expression and then solve the problem.

References

Carpenter, T. P., Fennema, E., & Franke, M. L. (1996). Cognitively guided instruction: A knowledge base for reform in primary mathematics instruction. *The Elementary School Journal, 97*(1), 3–20.

English, L. D. (1998). Children's problem-posing within formal and informal contexts. *Journal for Research in Mathematics Education, 29*(1), 83–106.

Fuson, K. C. (1992). Research on whole number addition and subtraction. In D. A. Grouws (Ed.), *Handbook of research on mathematics teaching and learning: A project of the National Council of Teachers of Mathematics* (pp. 243–275). Macmillan Publishing Co Inc.

Kilpatrick, J. (1978). Variables and methodologies in research on problem solving. In L. L. Hatfield (Ed.), *Mathematical problem solving* (pp. 7–20). ERIC/SMEAC.

Leung, S. S. (2009). Research efforts on probing students' conceptions in mathematics and reality: Structuring problems, solving problems, and justifying solutions. In L. Verschaffel, B. Greer, W. Van Dooren, & S. Mukhopadhyay (Eds.), *Words and worlds: Modelling verbal descriptions of situations* (pp. 213–225). Sense Publishers.

Ministry of Education of Taiwan. (2018). *Curriculum guidelines for the 12-year basic education elementary school, Junior High School, and Upper Secondary School. The Domain of Mathematics* (in Chinese).

Moser, J., & Carpenter, T. P. (1984). The acquisition of addition and subtraction concepts in grades one through three. *Journal for Research in Mathematics Education, 15*(3), 179–202.

Mullis, I. V. S., Martin, M. O., Foy, P., Kelly, D. L., & Fischbein, B. (2019). *HIGHLIGHTS TIMSS 2019 international results in mathematics and science.* https://timss2019.org/reports/

National Academy for Education Research. (2020). *Math curriculum guidebook* (in Chinese).

National Council of Teachers of Mathematics (NCTM). (1989). *Curriculum and evaluation standards for school mathematics.* National Council of Teachers of Mathematics.

Nesher, P., Greeno, J. G., & Riley, M. S. (1982). The development of semantic categories for addition and subtraction. *Educational Studies in Mathematics, 13*, 373–394.

OECD. (2023). *PISA 2022 results (volume I): The state of learning and equity in education.* PISA, OECD Publishing. https://doi.org/10.1787/53f23881-en

Polya, G. (1945). *How to solve it.* Princeton University Press.

Silver, E. A. (1994). On mathematical problem posing. *For the Learning of Mathematics, 14*(1), 19–28.

Chapter 3
Learning Grade Three Mathematics by Problem Posing for Communicating and Constructing Knowledge: Examples from Multiplication-and-Division

1 Introduction

This is a study conducted by an elementary school teacher, Kitty, who observed how students learn mathematics by using problem posing as a means for a whole year. At the time when this study was conducted, she was experiencing a period of curriculum reform in Taiwan. In the mathematics domain, the emphasis was on the active construction of concepts by the learners and not passively receiving the teacher's transmission of knowledge (MOE, 2008). The curricular document emphasized that the teacher designed materials and then presented tasks. After the presentation, the teacher did not demonstrate solving and left solving to students, who actively constructed math knowledge.

Kitty remarked, "*I see that children were not the same as compared to the children I taught 10 years ago. Um. I hope they can interact with math and can use math language fluently. They do not just memorize what I told them. As a class teacher who also teaches math, I can integrate the aim of respecting and accepting others' thinking in math teaching. On the other hand, I need to cope with this new wave of curriculum change. I must go for a graduate study to initiate this change.*"

Susan was a math teacher educator who taught at a local University in the same city where Kitty taught. She offered a bi-weekly Saturday teachers' workshop and shared research on solving and posing. Kitty was one of the 40 participants and impressed Susan with her in-class responses and frequently asked questions. Susan remembered Kitty's interests in math problem posing (MPP). They decided to work together on a project on its implementation. Kitty and Susan wanted to answer three questions:

S. S. Leung, *Mathematics Problem Posing in Action*,
https://doi.org/10.1007/978-981-96-8480-9_3

1. Among all chapters in grade three, were some chapters more suitable than others in using MPP instruction?
2. Take multiplication and division as an example, how did the teacher set up tasks to invite students to do MPP, and why?
3. What kinds of communication and construction of knowledge were found in students' outcomes?

2 Background Literature for Actions

2.1 Teaching to Meet National Curriculum Standards

At the time of the study, the document for Taiwan math curriculum standards was having revisions. In the previous document, it said "Students were to form a habit of actively constructing mathematics concepts, and also, through the understanding others' problem-solving processes, cultivate an attitude of respecting viewpoints of others" (see MOE, 2003); it emphasized constructivism. Content-wise, the strands were Number, Measurement, Shape and Space, and, Data Handling. In the next document, the General Guidelines of the nine-year curriculum, mathematics also includes the ability to apply knowledge and skills in their daily life; making the appropriate connections among materials and contents between this and other learning areas (MOE, 2003).

As above, both documents were influenced by the mathematics curriculum standards in the United States (NCTM, 1989, 2000), which emphasized problem-solving, communication, reasoning, and, connection. What happens if a teacher invites students to pose problems? When doing MPP the students needed to attend to problem structure (English, 2003). What if students pose different problems using the same task? The teacher can ask them to share what they pose, express themselves, and respect the different problems given by others. In this way, teaching using math problem posing will reach the curriculum standards (MOE, 2008). That is, to actively construct knowledge and to respect others' ideas.

2.2 Multiplication and Division in the Middle Grades

The teaching of number sense in earlier grades includes whole number learning: Counting, comparing sizes of two numbers, composition, and decomposition of numbers, addition, and subtraction. In the middle grades students learn multiplication and division. However, solving real-life problems regarding multiplication and division is different from knowing the two binary operations and doing a one-step computation on two numbers. Even before learning binary operations, children can solve a variety of multiplicative problems. They might not use multiplication and division operations. Research showed that they could use their solution strategies (e.g. repeated addition) to solve; and, they acquired various intuitive models of multiplication and division (Fischbein et al., 1985; Greer, 1992).

On whole numbers, there are four categories of problems on multiplication and division, according to semantic types (Greer, 1992; Mulligan & Mitchelmore, 1997).

1. Equal groups (e.g., 2 boys, each with 3 chocolates)
2. Multiplicative comparison (e.g., 4 times as many red as green)
3. Rectangular array (e.g., 5 rows of 6 flowers)
4. Cartesian product (e.g., 7 routes from home to school, 8 routes from school to city library)

The above four can lead to division problems when the total is known and one of the original givens is unknown. For example, equal group problems can be in the form of partitive division (e.g., finding how many chocolates each boy has). It can also be in the form of quotative division (e.g., finding the number of boys). Greer (1992) extended these categories to more types, involving when the concern is not only whole numbers but also fractions and decimals. As Mulligan and Mitchelmore (1997) commented, the categories can be extended, collapsed, or refined depending on the purpose of the study. For example, Greer (1992) gave an alternative conceptualization of an equal group, which becomes a rate problem. "If there are 3 chocolates per boy, how many do 2 boys have?".

When the multiplication or division problems were given in situations, and not a number operation, children read the situation and used their intuitive models to solve problems. Fischbein et al. (1985) explained implicit models. Each fundamental operation of arithmetic generally remains linked to an implicit, unconscious, and primitive intuitive model. Identification of the operation needed to solve a problem with two items of numerical data takes place not directly but as mediated by the model. The model imposes its constraints on the search process. Mulligan and Mitchelmore (1997) observed children solving the same set of standard word problems on multiplication and division. It was found that the students used three main intuitive models: Direct counting, repeated addition (or subtraction), and, multiplicative operation. All the intuitive models were used with all semantic structures.

When teachers use textbooks to teach, they follow the order of presenting word problems. It is customary that in the teaching of multiplication and division, word problems in textbooks are departmentalized (type by type) for practice, and, students often learn multiplication problems before division. However, in real-life contexts, the two kinds of problems are not separated and often come together. When students solve problems, they can use either multiplication or division; or even repeated addition or subtraction. If the teacher is not using the textbooks, she or he can design materials. If the teacher not only asks students to solve but also to pose problems, what types will students pose and solve? There is room for this investigation. When students do problem posing, will they pose problems with situations that call for a variety of implicit models, and lead to different ways of solving?

3 Method

3.1 Setting and Participants

3.1.1 Ms. Kitty and Her Belief

Kitty was an elementary school teacher who taught math for ten years. Her love for math education was unique. She signed up for teachers' professional development sessions for math subjects within and across districts. When she decided on a graduate thesis, she decided to study using problem-posing instruction to address two goals. The first was to *communicate math ideas and develop concepts.* Kitty believed that both the teacher and students needed to communicate math ideas. In problem posing, students were given opportunities to formulate problems from a given situation, or, to fix problematic problems to create legitimate problems. The second was to cultivate students' habit of *constructing math knowledge actively.* This approach was student-centered. Students were the main characters in class. The teacher was only a facilitator and guided students to develop their ideas yet respected peers' ideas. In her dreams, students love math, learn it well, speak math, organize ideas, and, not be afraid to present solutions in public. She liked to develop students' positive attitudes towards mathematics. The steps she followed represented a curious elementary school teacher trying MPP as one of the new ways of teaching given in the guidelines of the curriculum standards. In her class, she studied her children well and formed small groups. She liked to develop students' positive attitudes towards mathematics.

3.1.2 Students

The class size was thirty-five. They were divided into leaders, learners, and followers. The teacher divided the class into 7 groups; each group was composed of students from all three types (leaders, learners, and followers). Leaders were those who were usually subjective, eager to voice out their own opinions, could influence peers, and had high concepts of self. However, Kitty found that they were hard to correct when they made mistakes. Learners, the second type, represented those who listened to leaders and would not voice until they heard more from leaders and gradually gained confidence. Learners constituted the greatest number of students. The third type was followers; they represented the group who scored lowest in the class. They were those who originally wanted to give up learning math when they learned individually. However, when they stayed in small groups they interacted with learners and leaders and lost their feelings of isolation. As they worked in groups they got help and finally were not afraid of learning.

3.2 Design of MPP Program

A pilot task "The Snack Factory" for the initial use of MPP went well. "*Maybe not all chapters work the same way.*" Kitty and Susan then continued to work on materials for the whole year and compared them.

Materials. The investigators developed tasks by content and by MPP type. First, they referred to the math content according to the sequence given in the textbooks; a total of 9 units on Number Sense were included. Second, they made a table to display the type of MPP for each unit of the textbook. The six types of MPP were named according to what was given (see Leung, 2013). "*The aim is to teach mathematics and so we will consider the content and then types of MPP. The intended teaching materials will be used to decide on the MPP type to teach it.*" Kitty and Susan matched different pieces of math content to different types of MPP tasks. After the materials were developed, they met to talk about the steps in the program (see Fig. 1).

Steps in carrying out MPP. With both student-centered and critical thinking in mind, Kitty carried out the MPP instruction program and used 200 min per week. There were four lessons (40 min each), adding in two homeroom periods (20 min each).

STEP 1:
A double period (80 min) to start a new unit. The teacher presented problems for students to solve, then wrote math diaries-1. Later, they posed problems. When completed, they presented problems. Friends did query and the problem posers defended their problems. Students solved problems and completed diaries-2 (on posing) and diaries-3 (on personal experiences) in the lesson. Students handed in homework, and Kitty reviewed them and selected materials that were suitable for the whole class discussion during the next stage.

STEP 2:
One single period (40 min) to discuss the teacher's selection from students' diaries after STEP 1. The discussion included Kitty's selected mistakes/misconceptions, creative posing, and solutions processes worthy of discussion. The discussion was mainly for sharing, querying, and defense.

STEP 3:
A homeroom period (20 min) when students exchanged what they wrote from STEP 1 and STEP 2.

STEP 4:
Another lesson (40 min), referred to diaries from STEP 2 and exercises from STEP 3. Students learned to communicate their thinking and as well respect their peers' thoughts.

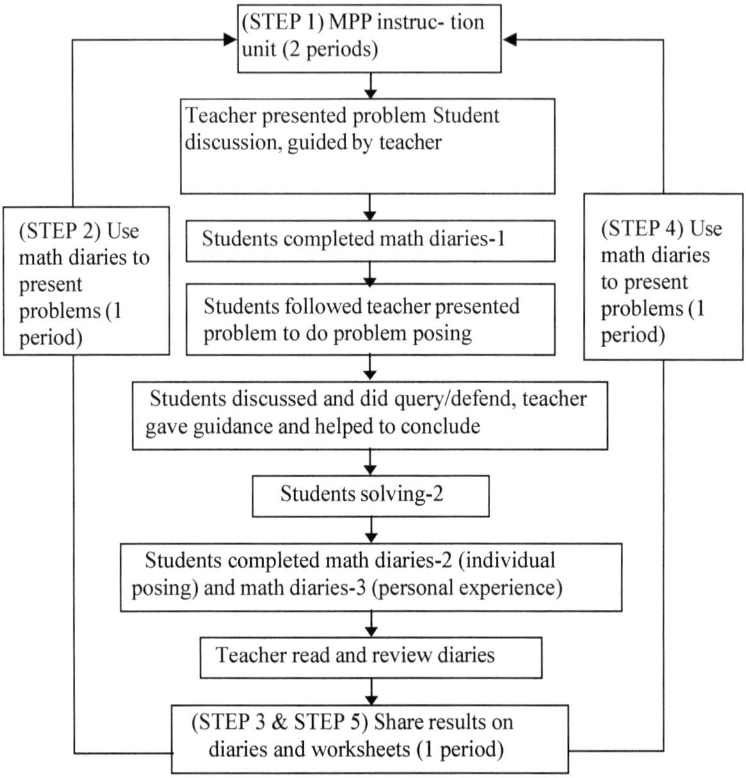

Fig. 1 MPP instruction program

STEP 5:
Another homeroom period (20 min) to appreciate the work from STEP 1, STEP 2, STEP 3 and STEP 4.

The above five steps form a complete cycle. Note that STEP 2 to STEP 5 could be repeated depending on the length of the unit.

3.3 Data Collection and Analysis

Data was collected from multiple sources: Math test scores and math exercise scores, teacher's journal (TJ), students' diaries (SD), video recordings (VR), and, observation notes of colleagues (ON).

Teaching's journals (TJ). TJ included Kitty's reflections on her lessons, enactment of MPP tasks and problems arising, her findings on students' learning outcomes, and what she learned from meetings with colleagues.

Students' diaries (SD). SD included three parts. First, record what students thought they learned. Second, the worksheet part with conditions, situations, and objects to pose a new problem and solve it. Third, students shared their feelings about the lesson and/or any personal needs.

Video recordings (VR). VR assisted the researcher in reviewing the lesson, what the teacher taught, and how the students behaved. Kitty called it a lesson review. There were three parts. First, an account of students' voices, distribution of voices of teachers and students, plus exchanges. Second, teachers monitored discussions on queries and key ideas. Third, episodes on students' responses to queries.

Observation notes (ON). A colleague who regularly visited Kitty's lesson wrote notes while he observed, mainly on students' responses that the video might not record.

There was a quantitative analysis of students' test scores. Besides, the investigators analyzed in-class data from MPP instructions qualitatively. Responses to MPP were coded and were analyzed qualitatively.

4 Results and Discussions

4.1 Comparison of Units for Students Learning Using MPP

Were some units more suitable than other units in using MPP? After one year of MPP instruction, math test scores and math exercise scores were used to compare Kitty's class to another grade three class teaching math without using MPP. Results are given in Table 1.

Among the nine units given in this table, five units made a significant difference: Multiplication, division-2 (division with remainders, also included division w/o remainders), multiplication and division, two-step four algorithms, and addition and subtraction on decimals. The first three units were related to multiplication and division. *Multiplication and division* will be used as an example to answer the two other research questions: Task setup, communication, and construction of knowledge.

Discussion on the comparison. Kitty and Susan wanted to answer the question, "Were some units more suitable than other units in using MPP to teach?" Kitty implemented MPP for a whole year in grade three and did a comparison among chapters from the textbook series. The investigators compared the characteristics of those units that made a difference to the other units that did not. The three characteristics emerged: *Extension, multiplicity,* and, *application.*

1. *Extension.* The units that made a difference allowed extension to other concepts and created cognitive conflicts for students (SD, VR). Students-teacher exchanges were non-stop; and, the contents of the exchanges were more in-depth and the atmosphere was extra lively (ON). When students did MPP they could extend ideas from one unit to another. For example, for the three units (multiplication, division with remainders, multiplication, and division) when students did

Table 1 Average scores for MPP teaching for the whole year (Grade 3)

Units	Math test			Math exercise		
	w/o MPP	with MPP	T-test	w/o MPP	with MPP	T-test
Three-Digit Addition	80.23	79.49	$t = 0.116$ $p = 0.751$	82.24	82.65	$t = 4.465$ $p = 0.102$
Multiplication	78.64	81.22	$t = 0.366$*	80.24	85.47	$t = 0.807$*
Fraction	82.08	82.36	$t = 0.105$ $p = 0.545$	84.62	84.66	$t = 0.024$ $p = 0.685$
Division-2	79.92	83.21	$t = 0.013$*	80.37	84.79	$t = 0.061$*
Vertical Addi-on	81.44	82.13	$t = 0.969$ $p = 0.381$	81.27	81.62	$t = 0.047$ $p = 0.639$
Decimals-1	82.53	82.34	$t = 0.007$ $p = 0.089$	82.04	82.64	$t = 1.001$ $p = 0.181$
Multiplication And Divisions	81.53	86.25	$t = 2.044$*	81.06	84.97	$t = 2.207$*
Decimals + −	83.62	85.47	$t = 0.153$*	81.97	85.76	$t = 0.255$*
two-step + − × /	79.97	83.66	$t = 0.907$*	79.34	84.82	$t = 1.182$*

*$p < 0.05$

problem posing in one unit (multiplication) and solved them they could automatically think of problems related to another unit (division or multiplication and division) (SD, TJ). For units like introducing fractions and decimals, there were lots of terminologies or declarative knowledge (*numerator, denominator, equivalent fractions, and, places after Decimals*) (TJ). Students found it hard to pose problems until they learned thoroughly the introduction (SD). In doing instruction on these topics, teachers needed to state definitions directly instead of asking students to construct them by completing MPP (TJ, VR). Later on, when students came across addition and subtraction of decimals they could extend ideas further (VR).

2. *Multiplicity*. In units with more than one single concept like two-step, four algorithms there was a variety of problems to go with this topic and the multiplicity allows a lot of room to create problems after solving one problem given by the teacher (SD). However, for units like three digits addition/subtraction, fraction, vertical addition/subtraction, and decimals were straightforward and emphasized computation and recording the results of a computation (ON, VR). For fractions and decimals, students remember the terms like numerators, or, places after decimals. For the two units on computations, the focus was on carrying out the steps in doing the operations. Therefore, the multiplicity of posing problems was not great (TJ, ON). These topics made MPP not easy; students found it hard to be creative in posing (TJ).
3. *Application*. For example, in two-step four algorithms and addition/subtraction with decimals these units allowed room for applications in solving word problems instead of merely making computations. Students when posing problems tried

very hard to make up problems that were not only different from the textbook but also different from their peers (SD). There were problematic problems that created conflicts among students and Kitty tried to orchestrate discourse until the whole class reached a consensus (TJ, VR). For the unit on three-digit addition/subtraction and vertical addition/subtraction, students merely learned about the rules of carrying and borrowing and also paid attention to place values (VR). Kitty already used storytelling to guide students to pose problems but most problems were very similar computation problems (SD).

As above, *extension, multiplicity,* and *application* marked the units when MPP was effective and appropriate. For the other chapters that emphasized computation or starting to introduce a brand-new concept, students were not as creative in MPP. If students were required to do MPP, they might end up just discussing for the sake of discussing, posing for the sake of posing. Kitty and Susan referred to the NCTM standards (NCTM, 1989) that suggested problem solving, communication, reasoning and connections. They found that a topic with connections also met the above three criteria for suitability of a unit in using MPP.

4.2 Design and Set Up a MPP Task: How and Why?

How did the teacher set up tasks to invite students to do MPP? and why? As given earlier, *multiplication and division* will be used as an example to answer this research question.

After the design, selection, and revision of MPP tasks, Kitty enacted tasks. She cultivated a culture of presentation of tasks, queries, response to queries, discussing, and reaching consensus. This way of teaching process encouraged active problem posing and solving of students. In the teaching, Kitty also emphasized on manipulation of real objects to help students concretize number concepts. Kitty constantly checked and recorded students' understanding or misunderstanding.

The following is a curriculum map of students' learning from grade one to grade three (from top to bottom). This curriculum map shows the order of teaching number sense topics including numbers, operations, and rules (Fig. 2).

Kitty referred to the order of learning sequence on numbers and found that the inter-relatedness among the topics. Among the materials for the whole year of MPP instruction, Kitty and Susan represented materials for three target units: Multiplication, Division-2, and, Multiplication and Division. Division-2 was the unit on Division with remainder, which also included division without remainder.

Multiplication: Set-up 1

1. Textbook's presentation

   A T-shirt costs \$120. How much should I pay if I buy three?

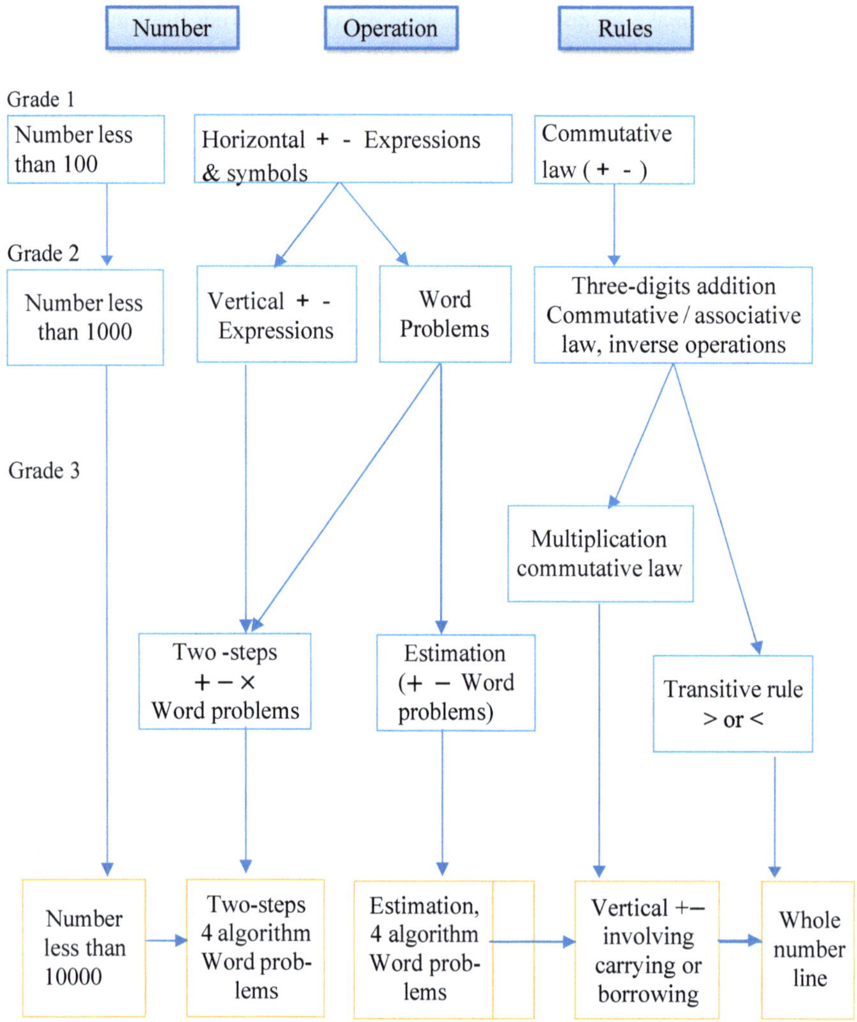

Fig. 2 Curriculum map of number sense topics

2. The teacher's presentation

   Candy factory

Kitty saw a candy house in the forest. She bought a box of Ferrero Rocher chocolate which cost her $120. She also bought 3 boxes of cool juice ($300 each), 4 Broad Bean Crisp ($25 each), and, 0 (zero) bag of Kuai Kuai potatoes chips ($10 each). There were other snacks in the Candy factory. Children, can you make a candy shopping *plan for Kitty? Then, calculate how much Kitty will spend, and bring your friends to the shop as soon as possible.*

Note: *Kuai Kuai* (in Chinese means "good child", usually used to describe those adorable child being pampered.)

How and Why Set-up 1: Zero bag of Kuai Kuai
Through designing a task called Candy Factory, Kitty guided students on commutative properties and in doing fill-in-blanks horizontal records on computations. She aimed for connections to previously taught ideas. About the numbers, Kitty purposely added a "zero" item. This was because she decided to arouse students' interest and be aware of using "0" and "1" (for a box) in multiplication. Also, to involve students in doing MPP, and to understand more about multiplicands and multipliers; also, to experience how to record vertical multiplication when solving. She hoped that students, in doing MPP, would refer to and revise prior concepts.

The textbook problem was a one-step problem, with three-digit numbers multiplying one-digit numbers. Students already learned the division of whole numbers and three-digit addition/subtraction. Therefore, Kitty planned to draw students' attention to the sizes of numbers regarding these two topics and posed problems about a creative purchase.

Division 2 (Division with remainder): Set-up 2
The tasks included both division with remainder and without remainder

1. Textbook's presentation
   There were 25 puddings, and three were packed in a box, how many boxes were there? How many remained?
2. Teacher's presentation
   In the Candy Factory house Kitty bought 29 pieces of chocolate. She decided to share with 9 students. If she paid each piece of chocolate $7, how much did she pay altogether? If she shared evenly, how many pieces of chocolate did each student get? Can you tell how you find the answer and why your answers are correct?

How and Why Set-up 2:
When students came to learn division 2 (also including division with remainder) they already learned multiplication so Kitty tried to revise it and link multiplication to division in a natural way. Adding the price of each piece of chocolate would get students to try do to both multiplication and division. Perhaps, students got the idea of inverse when they solved, posed, and solved problems using this CANDY factory situation.

According to the textbook problem, students could use 25 divided by 3, then find the quotient and the remainder (if any). This was a typical division problem. However, Kitty's problem set-up invited a variety of problems to be posed. Students also needed to spot-check the numbers to make sure the problem was solvable and there were answers. According to the textbook sequence, the part on checking answers to division with remainder was not required. It was not until students learned the two-step four algorithms word problems then the textbook problem required students to check answers by multiplication. However, Kitty's set-up made students pose problems and they learned to check answers even before they learned the two-step four algorithms.

Multiplication and Division: Set-up 3

1. Textbook's presentation
   There were 82 chrysanthemums, and there were 8 of each in the case. How many cases could be planted? How many chrysanthemums were there?
2. Teacher's presentation
   Kitty opened a big pet shop. There were 13 rabbits. Kitty wanted to buy the rabbits which were 6 times more than the original. There were 125 birds and 8 big birdcages. How can she put the birds?10 people wanted to buy dogs and dog food. However, Kitty's shop was still in a mess. Children, could you help Kitty to set up the pet shop?

How and Why Set-up 3:
"How to manage a pet shop" is a hot issue for students. Kitty believed that students could use this challenge to integrate division into multiplication.

Susan's note: Difference between textbook and teacher Kitty re- placed the straightforward stereotyped problems and hoped that to- wards the end of the school year students would wrap up ideas from both multiplication and division which would lead to two-step four basic algorithms.

4.3 *Communication and Construction of Knowledge*

What kinds of communication and construction of knowledge were found from students' outcomes? The investigators shared examples of multiplication, division-2 (also included division with remainder), and multiplication and division. Kitty let children present posed problems, discussed, made a query, responded and do defense, and try to help students to reach a consensus.

1. *Multiplication: How to buy watermelons*

Example of students' posing their problem:
We have to pose a problem using 6 × () = 42

I think it can be:
Fun went to a fruit stand to buy 6 watermelons and spent $42. How much is one watermelon?
Answer: 6 × (7) = 42, 7 watermelons

Kitty: Boys and girls. What do you think?
S32: Why use 6 × (7) = 42?
S22: Because we have to buy 6 watermelons for $42. A watermelon costs $7.

S5: I recite the Multiplication Table then I know.
S18: I also recite the Multiplication Table. However, it can be $7 \times 6 = 42$. Why it has to be $6 \times 7 = 42$?
S22: Don't argue. 6 is in front of the expression, so $6 \times 7 = 42$.
S: (don't believe)

Susan's note-1: Commutative Law
In the curriculum map, there was commutative law. Student S32 sent in a query, S22 answered using the context (watermelons, $7) but S5 and S18 use number facts from Multiplication Table they memorized. However, the students from the floor (S) showed that they did not believe. A teacher might use this opportunity to teach commutative laws ($6 \times 7 = 42$, $7 \times 6 = 42$). The part "it can be $7 \times 6 = 42$" created a cognitive conflict, the product was the same (42) but the way in grouping was different. However, Ms. Kitty let students discuss and communicate well on why and why not the same. She even asked if S22 was correct (*Don't argue*) as if she did not know. See below the debate continued.

Kitty: Is S22's answer correct? Does anyone have another explanation?
S18: (Raise hand)
Kitty: S18 please come to the stage.
S18: Because $42 \div 6 = 7$.
S: That is division not multiplication!
S18: Division and multiplication are reciprocal relations.
S27: But it has to match $6 \times (7) = 42$.
S18: It is checking calculation!

Susan's note 2: Checking calculation and inverse relation
Students from the floor did not accepted S18 because S18 changed the multiplication to a division. S18 defended by saying that the two were reciprocal relations. S27 sent in a query about matching the original "$6 \times (7) = 42$". The students were talking about the same relation being expressed differently. "Checking calculation" was S18's reply. This reply caught Ms. Kitty's attention. This inverse property (raised unintentionally through discussion) made S18 found convenient to do checking. Therefore. Kitty used a discourse strategy—REPEAT—to draw students' attention (See below).

Kitty: It is checking calculation. However, nobody tells me how to use $6 \times (7) = 42$ to buy watermelons.
S22: I know! (Face turned red, very excited)
Kitty: S22 please come to the stage and explain to everyone.

S22 came to the stage and drew circles on the blackboard (see Fig. 3)

S22: I draw 42 * which represents $42 and 6 circles which represent 6 watermelons (see Fig. 3). Every time I draw a * in every circle which means I spend $7 every time, for six times. Repeat this action until there are 42 * in the circle in total. I repeat. It means I spent $42.

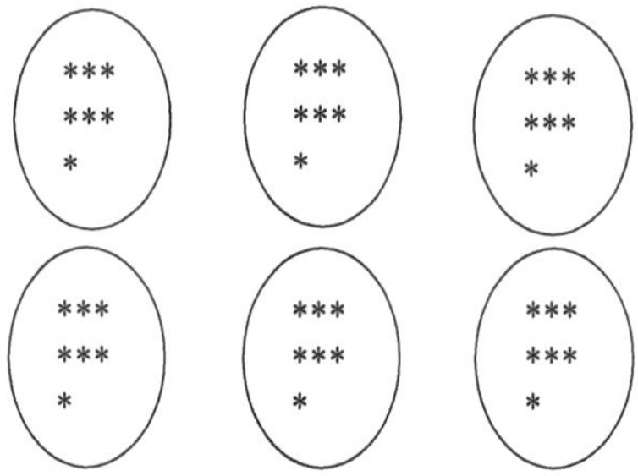

Fig. 3 Diagram of S22's solving on blackboard

Susan's note-3: Communication by drawing
S22 used drawings to remove 42 dollars. When S22 removed one for each melon S22 needed a dollars a time, and after 7 times, all 42 dollars were removed. Successive subtraction is a way of division (Mulligan & Mitchelmore, 1997).

The opportunities to learn appeared by chance during the debates.

1. Commutative law
2. Using a division to check a multiplication by division is using inverse relation
3. Drawing 42 and remove one by one can show repeated division.

When teachers require students to recite multiplication tables and insert "7", then, this problem can be completed quickly. However, inserting a correct answer without understanding the meaning was not the goal of teaching math.

Kitty let students present, query, defend, and discuss until they reach a consensus. "*Successful repetition and imitation without thinking the meaning is not the same as learning*" was the saying Susan and Kitty often mentioned during meetings.

2. *Division with remainder: Chocolates, gift packs and left overs*

Example of students' posing their problems

Kitty had 29 chocolates and she bought 8. How many chocolates were there in total? Kitty wanted to divide them into 6 Christmas gift packs. How many chocolates were in each pack? How many were left?

Answer: $29 + 8 = 37, 37 \div 6 = 6 \cdots 1$

A: (1) 37 (2) 6 (3) 1

S26: Why did you add and then divide?
S27: Because adding original chocolates to chocolates bought later meant *all* chocolates. Then we used division to divide into 6 gift packs. We can get the answer.
S26: Why can't we use division first? S27: I don't know what you mean.
S26: Means both used division!
Kitty: S26, can you write your idea and expression on the blackboard?
S26: I want to use division first just because the unit is about division. I don't know how to write an expression.

Kitty: The idea of S26 is great! You can try to record what you think! Does anyone can record S26's idea?

S15: I try! $29 \div 6 = 4 \cdots 5, 8 \div 6 = 1 \cdots 2, 4 + 1 = 5$ (tentatively, unsure) while S15 tried…S27 said below.

S27: It is not the same as my answer. You are wrong!

S15: I think that letting 29 chocolates be divided into 6 parts, and 8 be divided into 6 pacts is correct. Why do I have to add them first? Can't I divide them first then add?

Kitty: Is there any chocolate left after the division? What to do with the rest?

S3: Add them!

S15: How?

S3: Add the chocolates of the first left (5) and second left (2). It is 7.

S13: It is not the answer, too.

S3: If S27 is wrong?

Kitty: ok! Now we look back to the idea of S27. Whether you agree with this idea. Who can tell the idea of S27 to everyone?

S34: I try. Add 29 to 8 as 37 chocolates, then divide 37 chocolates into 6 gifts. There are 6 in every gift and 1 left.

Kitty: Great! Does everyone agree?

S: Yes!

Kitty: Is there anything wrong?

S: No!

Kitty: Ok, does anyone can tell S15's idea?

S15: I try! 29 chocolates divided into 6 packs, and in every pack, there are 4, 5 left; 8 chocolates are divided into 6 packs, and in every pack, there are 1, 2 left. Hence, there are 5 chocolates in one pack.

Kitty: How many chocolates are left after division?

S15: left are $5 + 2 = 7$.

Kitty: Can the leftover chocolates be divided into 6 packs?

S15: Ah! Yes! $7 \div 6 = 1 \cdots 1$. Every pack gets 1 and 1 left. Ah! Yes! Add 5 (original), there are 6 in a pack in total. I see!

Kitty: Can all of you record the expression of S15? Try!

All students try to record the expression:

$29 \div 6 = 4 \cdots 5$ (29 chocolates are divided into 6 packs, every pack gets 4, 5 left)
$8 \div 6 = 1 \cdots 2$ (8 chocolates are divided into 6 packs, every pack gets 1, then 2 left)
$4 + 1 = 5$ (every pack gets $4 + 1 = 5$)
$5 + 2 = 7$ (left $5 + 2 = 7$)
$7 \div 6 = 1 \cdots 1$ (the 7 are divided into 6, 1 left)
$5 + 1 = 6$ (there are 6 chocolates in every pack, 1 left)

Susan's Note: "Not the same" versus "Wrong"
The problem of division was done in two different ways. A student add up to a total then divide. Another student said "*Why do I have to add them first? Can't I divide*

them first then add?" Kitty did not insist on either one of the ways. This is because she accepted the solution as long as it was reasonable. She let the children discuss, communicate, and express ways clearly. When the student could not explain, Kitty did not explain for the student. She purposely asked another student to try to explain to check if the communication was well. After thorough discussion she asked students to make record of their work and not just by oral presentation. In turn, children learned not only math but also learned to respect others, even though the method was different. That friend's "Not the same" might not be "Wrong". After the discussion by presentation->query->defend->consensus the whole class found that both methods were correct.

3. *Multiplication and Division: Dog food*

Example of students' posing problems

Kitty and her 5 friends went to a pet store. The pet store sold guinea pigs in boxes of 2. A box is $24. How much does it cost to buy 3 boxes? A small pack of dog food is $22. The large pack is two times more than the price of the small pack. How much does it cost to buy two large packs? Kitty and friends add to a total of six people. How much does a person have to pay?

Answer: $24 \times 3 = 72, 72 \div 6 = 12, 22 \times 2 = 44$

$44 \div 6 = 7 \cdots 2, 12 + 7 = 19$

A: one person pays $19, but there are $2 that no one pays.

Discussion:

In the following lines of conversation, Kitty led the whole class to discussion. However, the part on "two times as much" and "two times more than" was difficult and confusing to grade students because of the language.

S29: Why nobody spends $2? Then how to pay?
Kitty: Do you agree with S5's solution?
S: Sounds strange. It seems not correct.
Kitty: S5, can you explain your solution to everyone?
S5: A box is $24, and it costs $72 to buy 3 boxes. 6 people share $72 and one person spends $12. A large pack of dog food is $22 × 2 = $44.
Kitty: what is the relation between a large pack and a small pack?

Please read the problem again!

A small pack of dog feed is $22. The price of a large pack is two times more than the price of a small pack.

Kitty: what's the meaning of "by two times"?
S: The price of a large pack is three times the price of a small pack.
Kitty So the price of the large is…?
S12: I know! The large pack of dog food is 22 × 3 = 66. A large pack is $66.

Kitty: According to S5's expression, $22 \times 2 = 44$, how to revise the problem?
S5: I know! The large is two times the price of the small rather than "by two times".
Kitty: Great, let's write down the expression. All classes try to record the expression.

$24 \times 3 = 72$ $72 \div 6 = 12$ (one person spends $12 on guinea pigs.)
$22 \times 3 = 66$ (one large dog food is $66)
$66 \times 2 = 132$ (two large dog food are $132)
$132 \div 6 = 22$ (six people share two large dog food)
$12 + 22 = 34$
A: Kitty and friend add to a total of 6 people, a person has to spend $34.

Susan's note: "by two times" means "three"
The fact that "*a large is two times more than a small*" was not clear to the student. The students thought that the large is two times as much as the small. However, Kitty did not say that the students were wrong. Kitty asked, "*What is the relation between a large pack and a small pack? Please read the problem again!*" This was her usual way to teach. Remind students to be careful instead of going to the stage and supply the correct answer, "*That meant 3 times*". Kitty still considered student-centered as her preferred way of teaching. She would prefer to invite other students to lead a revision, "How to revise the problem?".

5 Conclusion and Recommendation

A teacher worked with a teacher educator and found that teaching via MPP is feasible when the teacher carefully select a topic, design and enact tasks, orchestrate discourse to communicate and construct knowledge. First, after a year of designing, and using materials for MPP instruction for grade three, few units made statistical differences in test scores when compared to a class that did not use MPP in teaching. They have three characteristics: *extension, multiplicity*, and, *application*. Second is about design and set up of tasks. Since the teacher could not tell what students posed and when they made mistakes the preparation for MPP was most difficult, no rehearsal at all. To her, it was very challenging indeed. The preparation of materials was found to be important in that students had opportunities to pose a variety of math problems. The third is on how knowledge can be communicated and constructed by students. The process was cyclic and dynamic. The teacher presented problems > students solved problems > students posed problems > peers sent queries > the whole class discussed for consensus. Students' communication was encouraged and they showed interest in posing and solving, sharing work, and learn from others, as well as respecting the thoughts of peers, which coincided with what was advocated in the curriculum reform standards (MOE, 2003, 2008).

The use of MPP as a means for students to actively learn math is given in this chapter. Kitty tried all chapters in grade three. The example here is multiplication and division. As earlier work showed, this topic "multiplication and division" is rich and allows room for investigation into implicit models (Fischbein et al., 1985; Greer, 1992). In problem posing, a variety of problems on multiplication and division can be resulted, in different categories according to semantic types (Greer, 1992; Mulligan & Mitchelmore, 1997). However, all other topics can be explored in future for suitability and for other grades under different situations. Task, subject and situation are variables in research on problem solving (Kilpatrick, 1978) and will also be investigated in problem posing, as given in the forthcoming chapters. Please also see Fig. 3 again from chapter one of this book).

References

English, L. D. (2003). Engaging students in problem posing in an inquiry-oriented mathematics classroom. In F. Lester & R. Charles (Eds.), *Teaching mathematics through problem solving: prekindergarten-grade 6* (pp.187–198). National Council of Teachers of Mathematics.

Fischbein, E., Deri, M., Nello, M. S., & Merino, M. S. (1985). The role of implicit models in solving verbal problems in multiplication and division. *Journal for Research in Mathematics Education, 16*(1), 3–17.

Greer, B. (1992). Multiplication and division as models of situations. In D. A. Grouws (Ed.), *Handbook of research on mathematics teaching and learning: A project of the National Council of Teachers of Mathematics* (pp. 276–295). Macmillan Publishing Co Inc.

Kilpatrick, J. (1978). Variables and methodologies in research on problem solving., in L. L. Hatfield (Ed.). *Mathematical problem solving* (pp. 7–20). ERIC/SMEAC.

Leung, S. S. (2013). Teachers implementing mathematical problem posing in classroom: Challenges and strategies. *Educational Studies in Mathematics, 83*, 103. https://doi.org/10.1007/10649-12-9436-4

Ministry of Education (MOE). (2003). *The MOE deep cultivation project for nine-year curriculum and instruction.* http://edu.law.moe.gov.tw/LawContentDetails.aspx?id=FL029037&KeyWordHL= (in Chinese).

Ministry of Education of Taiwan. (2008). *General guidelines of grades 1–9 curriculum for elementary and junior high school education* (in Chinese).

Mulligan, J. T., & Mitchelmore, M. C. (1997). Young children's intuitive models of multiplication and division. *Journal for Research in Mathematics Education, 28*(3), 309–330.

National Council of Teachers of Mathematics (NCTM). 1989). *Curriculum and evaluation standards for school mathematics.* National Council of Teachers of Mathematics.

National Council of Teachers of Mathematics (NCTM). (2000). *Principles & standards for school mathematics.* National Council of Teachers of Mathematics.

Chapter 4
Learning Addition and Subtraction of Fractions and Decimals Using Problem Posing: Two Case Studies

1 Introduction

According to Taiwan mathematics curriculum standards (MOE, 2008), the four strands in this subject domain were Number and mensuration, Space, Data handling, and Algebra. Number and mensuration constituted an important and major part at elementary levels. In the learning sequence of numbers, the whole number goes first, and then the learning of rational numbers follows. However, the learning of rational numbers is always a challenging topic. The learning sequences among the many regions in the world are two. The first curriculum sequence is whole-fraction-decimal, and the second is whole-decimal-fraction. No matter what the sequence is, there was a possibility of errors in number learning because children used what they previously learned to make sense of new instructions. (e.g., Resnick et al., 1989). For the first sequence, children may treat fractions as whole numbers and add two fractions together by finding the sum of two numerators, even though the two denominators are different. For the second sequence, they may treat decimals as whole numbers, and align two decimals (e.g., 0.1, 0.234) to the right then do vertical addition to get "0.235" as the answer.

In Taiwan, the teaching sequence is the first one: whole-fraction-decimal. The teaching of fractions precedes that of decimals. When teaching fraction numbers (which is abstract) we can use explanations on quantities (which is concrete). When dividing a quantity, the names of quantities change from a whole to parts, no matter whether the quantity is discrete or continuous. When representing a number as a fraction, students learn terminologies like *numerator*, *fraction bar*, and, *denominator*. When representing a number as a decimal, the dot between the whole number and the fractional part is called the *decimal point*. Students also learn to count the number of *places after the decimal point*. However, the part on operations on rational numbers is very different from that of whole numbers.

S. S. Leung, *Mathematics Problem Posing in Action*,
https://doi.org/10.1007/978-981-96-8480-9_4

Susan, a math teacher educator worked with two teachers, Mr. Please and Mr. Stream, and attempted to teach rational numbers using math problem-posing instruction. Mr. Please taught in the east of Taiwan. Students were from aboriginal families. The target was on fractions, including multiplication and division of whole numbers, addition and subtraction of two fractions with different denominators, and comparing two fractions with different denominators. Mr. Stream taught in a city in south Taiwan. He and his students were having a transition period under curriculum change. His target was on decimals, including comparing two decimals and decimal addition and subtraction. This chapter concentrates on the common part of the two teachers' attempts: the addition and subtraction of two numbers (fractions, decimals).

Below is how two teachers, Please and Stream, completed two different but equally challenging journeys with a math teacher educator (Susan).

The research questions are: (1) What were the changes to students' learning when MPP was used to teach addition and subtraction of two rational numbers? (2) What affective (acceptance) and cognitive behaviors (conceptions and misconceptions) could be found when students through problem posing and problem solving word problems? (3) What did the teachers suggest on how students learn rational numbers using MPP?

2 Background Literature for Actions

The following included what Susan shared with Please and Stream, the background for carrying out a study on fractions and decimals.

2.1 Rational-Number Concepts: Fractions and Decimals

The word "fraction" stems from the Latin word "frangere" which means break into pieces (Frobisher et al., 2002). The meaning changes according to the contexts and how the "break-into-pieces" occurs. The word "decimal", checked from the Oxford dictionary, is a fraction whose denominator is a power of ten and whose numerator is expressed by figures placed to the right of a decimal point. Children come across these two topics in elementary school, after they learn whole numbers. Behr et al. (1983) discussed the acquisition of mathematics concepts and processes in rational numbers. They considered rational number concepts as the most complex and important mathematical ideas children learn in primary school. Three perspectives were given: children need to also deal with numbers that are not only whole numbers in real-world situations; children further develop mental structures necessary for continued intellectual development; and, rational-number understandings provide children with the foundation upon which elementary algebraic operations can later be based. The interpretation of rational numbers can be given in sub-constructs (Vergnaud, 1979) such as a decimal, a ratio, an indicated division (quotient), an operator, and a measure of quantities (continuous or discrete). The understanding of each of these separate sub-constructs and, how they interrelate is important in the acquisition of rational number concepts. In studying problem posing involving rational numbers, the above perspectives can be used, to analyze the understanding of separate sub-constructs and how they interrelate.

2.2 *The Learning of Fractions and Decimals Under the Climate of Curriculum Reform*

According to local curriculum standards (MOE, 2003), students learn whole numbers, fractions, then decimals. As far as the idea of fractions is concerned, and in the context of splitting "quantity", it is represented in the form of juxtaposing two numbers (one on top of the other), separated by a horizontal line in the middle. Students learning of integers may interfere with thoughts, resulting in a concept and format algorithm error. For example, it is hard for students to distinguish part/whole, subset/superset, or fractions on a number line, or which part represents a unit measure (Behr et al., 1984; Cramer et al., 2002).

Mr. Please found that operations on fractions are tough. When teaching whole numbers, addition and subtraction were taught before multiplication and division. However, for fractions, addition and subtraction were taught after multiplication and division; and, such operations cannot be done in one step. *"Maybe students need to experience fractions in a variety of word problems. If they pose and solve word problems the problems will make sense to them. This could help them to bridge the math world to the real world."* In his prior teaching on the part on mixed fractions and improper fractions, his students found it very hard to imagine in real-world contexts.

Another teacher, Mr. Stream, also found that operations on decimals were tough. The decimal number is considered as one particular type of fraction. Mr. Stream said, "*I like to know what will happen when I conduct MPP instruction for decimals to grade six class, how they problem solve, how they problem pose and what misconception(s) appeared (if any).*" He also remembered that when he taught the part on different places after decimals, students found it hard to imagine and compare the sizes. "*I will try posing word problems. In the math world, students read and said 0.199 is greater than 0.2. However, in the real world, students know 0.199 kg beef is less than 0.2 kg beef.*"

Susan emphasized the importance of units in performing addition and subtraction. For whole numbers, say "3" plus "4", means getting how many ones in all (a total of "7" ones). For multi-digit numbers, operations can be done directly by alignment of two multi-digit numbers. In whole number operation, the unit is ones. However, when considering operations on rational numbers, the steps may involve converting one number (Fraction, decimal) to another equivalent form of number having the same unit. The unit is not necessarily one. For fractions, when adding a half to a third the two fractions are often converted to sixths ("3" sixths to "2" sixths). After changing to equivalent fractions with a common denominator, the unit is then one-sixth, and the sum is "5" sixths. For decimals, adding 0.2 and 0.199 means adding 200 thousandths and 199 thousandths; the unit is then a thousandth. No matter what the number is (whole numbers, fractions, and decimals), the operations must be carried out by choosing a convenient unit in common. She provided this background to the two elementary school teachers (Please, Stream).

3 Method

Susan worked with two teachers and met to solve problems in studies on MPP. Mr. Please and Mr. Stream, each had a plan of study and carefully considered participants, teaching materials, and data collection and analysis.

3.1 Participants

3.1.1 Case of Fractions

The teacher was Mr. Please. The teacher was a veteran teacher who taught in the east of this island with students mainly from aboriginal families. At the time when this study was carried out, he had been teaching in this school for seven years. He knew students very well.

Over the half a year of daily interactions, they even met for supervision class after school each day. Mr. Please had his training from four seminars in math education from Susan during graduate school. The proposed teaching idea, using MPP, was a new attempt to him, and very few researchers worked on MPP teaching for aboriginal students.

Students of Please were twelve grade 5 elementary school children (four boys and eight girls). All of them are from aboriginal families (10 Pai-wan, 2 Amei). The location of the school was in a quiet village. Parents were mostly farmers and workers and rarely committed to business. In this village, there was quite a proportion of social gentry and families working in public agencies, government officials and representatives of public opinion in the community, and medical doctors. Although parents showed concern for children's learning, their efforts were more focused on school events like sports day and not on academic subjects such as math. Parents seldom interacted with teachers. They might call to thank teachers for checking on students' work and progress.

3.1.2 Case of Decimals

The teacher is Mr. Stream. He taught elementary school for over 10 years and wanted to upgrade himself by attending graduate school. He had his training from four seminars in math education from Susan during graduate school. When he came across problem posing, he was excited about it and used it in instruction. Finally, he decided to study the integration of MPP, how students solved it, how students posed, and if any misconceptions appeared during the lessons.

The students were from a sixth-grade class of a primary school in south Taiwan, there were 16 boys and 16 girls, a total of 32 people. At the time of the study, students in this school used 2 mathematics textbooks, the 82nd edition for the first to third grades, and the nine-year provisional edition for the fourth to sixth grades. Therefore, there was a need to cater for a transition from one curriculum standard to the other.

3.2 Design of MPP Instruction Materials

3.2.1 Case of Fractions (Please's Class)

Mr. Please found that operations on fractions are tough. Unlike whole numbers, addition and subtraction were taught after multiplication and division; and, such operations cannot be done by one step.

The kind of MPP here means solving one particular type of fraction problem and then posing one problem that is either similar or different. After the problem is posed, problem solving follows. Mr. Please read to find that among the chapters in the fifth-grade textbook, his students found fractions very difficult. "*Fraction is always a difficult piece in the elementary mathematics curriculum. In posing and solving problems on fractions, what can students learn if I add MPP? Are there changes in students' learning of fractions? How about their attitude? And their acceptance level of MPP? What will I include in journals for teachers' reflections?*".

Mr. Please designed MPP worksheets and test items following the given contents respectively. There were four times per week and two weeks per unit. To further discuss results with that of Mr. Stream, only the unit on addition and subtraction was reported in this chapter.

Mr. Please's typical class included two phases.

<u>Phase 1 (40 mins): regular class daytime</u>

1. Revision on fraction concepts (3 min)
2. Teaching and exercise (5 min)
3. The teacher posed a problem then students solved problems (7 min)
4. Teacher clarified concepts (5 min)
5. Students posed problem (20 min)

   5-1. Problems legitimate, invited students to solve

   5-2. Problems problematic, clarified further, revised presented, and discussed.

Phase 2 (40mins): extra supervision class after school

Clarify/consolidate concepts, students wrote results on posing and solving (10 mins)

Students completed the exercise (20 mins)

The teacher graded exercise and assisted in revisions (10 mins) Students completed diaries at home and shared them with peers and teachers the next day.

3.2.2 Case of Decimals (Stream's Class)

Mr. Stream provided word problems on *addition or subtraction, carrying/borrowing or no carrying/no borrowing*, and, *single-step or two-step* (+ – ; – +). After these combinations would add up to eight items. He also made sure that all numbers have three (or fewer) places after decimals.

Below is an item with addition/no carrying/single-step.

Red Fire Ants carried cookies for winter. The first time they moved 2.325kg, and the second time they moved 1.65kg. How many kilos did they move in total?

Ms. Stream's typical MPP class

1. The teacher presented one problem and students solved it.
2. Students posed problems and solved their posed problems
3. Students shared posed problems and wrote diaries
4. The teacher completed journals and recorded his teaching and what he found in students' diaries.

3.3 Data Collection and Analysis

3.3.1 Case of Fractions (Please)

Mr. Please collected data including fraction tests (pre, post), MPP worksheets, students' diaries, teacher's journals, and LIKE sheets.

1. Math tests on fractions (Pre and Post)

   There were altogether two tests, which corresponded to the two chosen units on fractions. Students took a pre-test before instruction and a post-test after each unit. The test items were given in Appendix 1, a total of eight types.
2. MPP worksheets

   A worksheet on one type is given below. (See Appendix 1 for other types).

   e.g. Type 1 (Addition, Two fractions with one denominator as multiple of the other)

   "Dad bought $\frac{2}{5}$ kilograms of pork and $\frac{2}{20}$ kilograms of beef to cook a big dinner. How many kilograms of meat did Dad buy in total?"

   Try yourself. Can you solve the problem you make? Please write down how you find the answer.

If you were Mr. Please, can you think of a similar or different problem? Please write below.

Self-evaluation: __Very good __Good __Quite good __Try harder

Mr. Please's: __Very good __Good __Quite good __Try harder

3. Students' diaries

Students completed at the end of the unit.

It was a communication bridge for Mr. Please to reach students. It included: *my creative problem, my solution, and, math from my heart*. Students used words, numbers, drawings, and other written forms to express what they thought about the lesson.

4. Teacher's journal

In his journal, Mr. Please kept a record of his thoughts before, during, and after each lesson. He divided his journal accordingly: *Anticipating, observing, and reflecting.*

5. LIKE sheets

Students completed at the end of the unit

Do you like the way we had math lessons? Why? (at least 30 words)

1. Do you think posing your problems is fun? Do you think you improve in mathematics learning? Why? (at least 30 words)
2. Do you think writing math diaries helps you to learn math? Why? (at least 50 words)

3.3.2 Case of Decimals (Stream)

Mr. Stream collected data including problem-solving worksheets, problem-posing worksheets, students' diaries, teacher's journals, and, audio-tapes of lessons.

1 Problem Solving worksheet

Problem (read below):

Please write an expression for solving the above problem Solve this problem and find the correct answer.

2. Problem Posing worksheet

After you solve the problem, please make up another problem Please write an expression for solving the problem you just made Solve your posed problem and find the correct answer.

3. Students' diaries

What did I learn from today's lesson?

About today's lesson, what did you feel? (please write or draw how you feel)

4. Teachers' journal

Teaching progress

Students' responses and feedback Reflections

5. Audio-tapes

Lessons were audio-taped.

As above, the five sources of data were collected by Please and Stream; and, the team of investigators used mixed methods (quantitative and qualitative) to do the analysis. Coding of the solvability of the posed problems was done by following the five types as in Leung and Silver (1997).

4 Results and Discussions

4.1 Changes to Students' Learning After MPP Was Used to Teach the Addition and subtraction of Two Rational Numbers.

4.1.1 Case of Fraction (Mr. Please)

Quantitative measures. There was a total of 192 worksheets (12 students, 8 items, on two units). After comparing math test scores (pre, and post), Mr. Please used descriptive statistics to check if there were changes in fraction test scores. Students performed better in the post-tests.

Qualitative results. Examples were taken from the unit on addition/subtraction of two fractions with different denominators. Students' outcomes on MPP might be unexpected, incomplete, or unclear wording; each case led to interesting discussions. Three episodes were selected: *Common denominator, sameness, and, aboriginal features.*

1. (C*ommon denominator*) The first story was on two fractions with different denominators, one can be the multiple of the other, versus, two that were co-prime. The different cases required different numbers of conversions to have a common denominator. Mr. Please asked students to solve one problem involving two denominators with one being a multiple of the other.

Mr. Please: *Dad bought $\frac{2}{5}$ kg pork and $\frac{3}{20}$ kg beef to prepare a great big feast for dinner. How many kg of meat altogether?*

He invited them to solve and they changed the first fraction to $\frac{8}{20}$.

The computation was done smoothly. After students found solutions,

Mr. Please asked students to pose a similar problem but he forgot to specify the requirements for the two fractions, that is, with one denominator as a multiple of the other.

S11 posed a problem that asked for the lengths of ribbons.

S11: *Sun bought two gifts and needed ribbon for wrapping. One required $\frac{4}{5}$ m and the other $\frac{2}{3}$ m long. How much ribbon did Sun use?*

However, in this ribbon problem, the two denominators were co-prime, not as Mr. Please expected: Involving two denominators with one being a multiple of the other.

Upon reading work of S11, Mr. Please asked S11 to solve. Unlike the meat problem (2/5 and 3/20), where 5 can be changed to 20 using one step, thirds could not be changed to fifths with one step of conversion. After some struggle and failure to convert fifths to thirds, S11 did two conversions and used "15" as the common denominator. S11 was excited to succeed after making two conversions.

$$\frac{2}{3}+\frac{4}{5}=\frac{10}{15}+\frac{12}{15}=\frac{22}{15}=1\frac{7}{15}$$

As above, a student might pose a problem not the same as the teacher but that allows room for investigations into fractions addition.

2. (*Sameness*) The second story was on units or "same" or "not the same", as in children's words. This story explained how problem posing made students learn the importance of units in operations such as adding or subtracting.

S07: Windy ate $\frac{1}{4}$ guava and $\frac{5}{10}$ apple, how many fruits did she eat?

Her friends challenged her and one of them asked her, "I cannot add guava to apple, is the guava the same as the apple?" Later, they developed the discussion to the definition of "same" and more students joined in the discussion.

Instead of teaching them to revise the posed problem, how Mr. Please stayed quiet.

> *"This is interesting. I better let them talk and give their own opinions on the same... Let me pretend that I did not notice and said nothing until they agreed on something they could agree. I better stay quiet."*

One student suggested changing apple to guava. Another changed all fruits to apples. The preferred "same" to be the same kind of fruit. However, the children then complained that not all apples nor guava were the same. This was because one student added, "*I cannot add apple to apple, one apple could be large and the other small*". Hence, a change to the same kind of fruit could not help. The students wondered, *"Same in what sense?"*

Finally, student S07 came to into rescue.

S07 was the person who posed this problem. He said he did not mean a fractional part of a fruit. "*Sorry, I was careless and omitted kilogram, I meant I forgot the units*". He revised his posed problem after friends questioned him and challenged him on sameness. He added "kg" next to the numbers to clarify the quantity of fruits.

S07: Windy ate $\frac{1}{4}$ kg guava and $\frac{5}{10}$ kg apple, how many fruits did she eat?

It was not until his friend read the revised problem by S07, they did not realize that S07 meant weight of fruits and not a fractional part of the fruit. After they agreed on the problem they solved it, and one of them put up the solution to the blackboard.

$$\frac{1}{4}+\frac{5}{10}=\frac{5}{20}+\frac{10}{20}=\frac{15}{20}$$

A friend asked if the answer (15/20, a fraction) was better when in simplest form and they all changed the answer to "3/4".

Another example was on units for measurements.

S06: S6 posed a problem on areas and he used cm. Upon presentation, S6 was asked by friends to revise measurement units for area, so, S6 changed the measure in length (cm) to a measure in area (cm square).

3. (*Aboriginal features*) The third story was about aboriginal features given in the word problems.

S10: On *two identical pieces of land Dad planted lilies on* $\frac{3}{6}$ *of the first piece and Mom planted lilies on* $\frac{3}{4}$ *of the second piece. Which 4 parent-planted piece was larger?*

The student, S10, who posed this question was an aboriginal girl. Lily was a flower that was very popular in this village and S10 posed problems in planting lilies. In the tribe, this flower represented bravery (for male) and purity (for female).

S05: *Amy danced for* $1\frac{4}{6}$ *hrs. and then, sang for* $1\frac{4}{6}$ *hrs. How many hours did she spend on dancing and singing?*

S05 (also S10) were two girls who loved to sing and dance which were typical leisure activities for aboriginal children. It is well known that aboriginal girls are good at singing and dancing.

S12. Mr. Please presented a problem involving a standard box of cola with 12 bottles. He was expecting students to use bottles and boxes to pose problems using fractions. After solving S12 posed and changed names to aborigines (friends' names), and used different numbers. Cola was changed to rice wine, which was more favorite than Coke cola to the Aborigines.

S03. It was hard for Please to find real situations using fractions. He was thinking of using weights. "*Let me use weights, which are not necessarily in whole numbers.*" Please struggled with the choices and finally used 7/11 kg as *the* weight of a pot. Students posed similar problems. S03 posed and changed to weights of boar which was easily found in the near region of students in this school. Boars were familiar to Aboriginal children.

Susan's note for discussion

On students' learning of fraction addition (Please's students)

The first is student-centered. Usually, when the teacher used one from the set of problems exercise, all the other problems from the set were of the same kind: one denominator is a multiple of the other denominator (Type 1). In posing and not solving another problem from the set, an unexpected problem was posed by students. However, student S11, still completed it, with two "one-step" problems. The teacher respected this problem and did not asl the students to pose again. Please's classroom is student-centered as suggested by the curriculum standards (MOE, 2003). He let S11 try instead of changing the problem back to one he preferred.

The second is the importance of units: That "when units are not the same, you cannot add" was emphasized over the meetings for the three investigators. Mr. Please patiently waited for the children to communicate their thoughts and came to a consensus. Children learned the importance of having a unit. From the story on fruits, the teacher's decision was not to interrupt students' discussion but to allow them to talk. They asked questions about sameness. That "same" referred to the kind of fruits, then, the "same" meant size, and finally, the "same" turned to measures such as weights. Mr. Please was also happy to see that the reminder, in simplest form, was put forward by a student instead of by him. When adding, the units have to be the same (kg). From the story on areas, the problem is with using square centimeters as the units.

The third is about contexts that are meaningful to students. When students posed and solved problems, they used contexts they found interesting. The above aboriginal examples were: the flower Lily (S10), Singing and dancing (S05), Rice wine (S12), and Boar (S03). Aboriginal females loved Lily, which symbolized the purity and kindness of Paiwan and Rukai women. Aborigines love to sing and dance, especially during their traditional ceremonies or gatherings. Also, home-brewed millet wine was often a drink to add fun, and wild boars captured by hunting were delicacies shared by the family. S03 and S12 are the little warriors of the aborigines. The two students remembered common behaviors of adults. Mathematics problem posing was then very life-oriented and an indispensable activity in daily life.

4.1.2 Case of Decimal (Mr. Stream)

There was a total of 287 posed problems, and there were three types from the five given by Leung and Silver (1997).

1. <u>Problem with sufficient data (n=278; 96.9%)</u>

   A bamboo pole 3.55m long. The snail started to climb from 1.38m above the ground. After climbing 1.29m, it met dew and slipped 0.37m down. How many meters above the ground was the snail?

   Note: The students used analog to pose the problem. Mr. Stream invited students to solve as this problem was ready for solving.
2. <u>Problem with insufficient data (n=4; 1.4%)</u>

   Fish loves to play video games and spent 4.87 hours playing, and, at noon Smart played with fish for 1.84 hours. How much time did they spend in playing? (7-p-12)

   Note: It was unclear if 4.87 included the time Fish played with Smart or not. Mr. Stream invited students to revise the solution.
3. <u>Implausible problem (n-5; 1.7%)</u>

   Ming first took 3.2 kg and then took 1.45 cm. How many altogether? (2-p-14)

   Note: the problem was posed by a student with low academic achievement. Mr. Stream followed with an interview and asked him to clarify the work. After the individual interview, the students realized that just adding two different measures, such as weight and length, was meaningless.

The differences in word problems of students posed problems to Stream's presented problems were noted. It could be in structure, objects, or numbers. By structure, it meant changing the given and the unknowns. Among the 287 posed problems, Stream counted the number of changes in structure (16; 6%), objects (276; 96%), and numbers (287; 100%).

Finally, Mr. Stream also compared the percentage of success in solving his problem to the percentage of success in solving the student's own posed problem.

As given in Table 1, the rate of success in solving posed problems was higher than solving the teacher's given problems (except 2nd). The same pattern of numbers change applied to the comparison for misconceptions and miscalculations. Also, MPP could help students to clarify the misconceptions and understand more about decimals, and the course content. In terms of problem-posing teaching, it could enhance learning interest, correct misconceptions, which responded to the suggestions given by the curriculum standards (MOE, 2008).

Susan's note for discussion

Mr. Stream used results in the teaching of decimals to study students' change in learning. They did better in solving their posed problems and made fewer mistakes. Mr. Stream made use of students' posed problems to teach math concepts and addressed problems or misconceptions indicated in students' work accordingly. When students' posed problems were sufficient, Stream replaced textbook problems with students' posed problems.

When problems were insufficient to solve Stream invited them to be engineers and "fix" the problem. In fixing the problem into one that could be solved, students also learned why the original ones were insufficient.

4.2 Affective and Cognitive Behaviors

4.2.1 Affective Behaviors (from Please's Study)

The investigators analyzed the LIKE sheets to understand students' acceptance of mathematical problem posing (MPP).

LIKE. Most of them like (9 out of 12). I like this way of teaching because it can improve my math and make my mind more flexible (S02). It's not bad, because it can improve my math, and it can also help me in daily life. It will be of great help when we go to senior high school or junior high school (S07).

There were 3 who checked DISLIKE. I don't like this way of teaching math. Math is very difficult for me already. Completing the math posing sheet is also difficult for me…(S09).

They also considered if MPP was FUN from the worksheet. *Very good, because you can pose very simple problems, and more difficult problems to test yourself. Then you improve in math (S03) I like it because when I did problem posing, I was both a teacher and a child and teaching myself to pose problems. I can experience the fun of finding questions and solving them by myself* (S07). Only one said it was not FUN.

It's not fun. Sometimes too difficult. You do not know what to pose, and you will feel very bad… (S05)

Table 1 Students' success in problem solving and problem posing

Order	Name of problem	Solution type	Success in solving ($n = 32$)	Reasons for not success in solving	
				misconception	miscalculation
1st	Xiao Xin Long Jump problem Weight Comparison problem	Solving problem	27 (84%)	5	0
		Solving posed problem	29 (92%)	3	0
2nd	Red Fire Ant problem	Solving problem	30 (94%)	1	1
		Solving posed problem	29 (92%)	0	3
3rd	Orange problem	Solving problem	27 (84%)	0	5
		Solving posed problem	31 (97%)	0	1
4th	Rumi goes to school problem	Solving problem	31 (97%)	0	1
		Solving posed problem	31 (97%)	0	1
5th	Papaya Milk problem	Solving problem	28 (87%)	2	2
		Solving posed problem	30 (94%)	1	1
6th	Snail Climbing problem	Solving problem	24 (75%)	6	2
		Solving posed problem	28 (87%)	2	2
7th	Pippi Cat Drinking Milk problem	Solving problem	25 (79%)	2	5
		Solving posed problem	27 (84%)	2	3
8th	Conan's eating problem	Solving problem	28 (87%)	2	2
		Solving posed problem	31 (97%)	1	0
9th	Pork Dumpling problem	Solving problem	27 (87%)	1	3
		Solving posed problem	29 (94%)	1	1

WRITING. Undoubtedly, almost all students agree that math notes help learn math (only one did not). They expressed positively about writing math notes. *I think the math notes help me a lot because I can pose my problems and solve them by myself. I hope there will always be a math-posing sheet that cultivates our knowledge* (S08).

FRIENDLY. In doing MPP and in exchanging problems to solve, students learned to be friendly while doing problem posing. Students did not want friends to have headaches in solving math problems. Among the posed problems by twelve students, Mr. Please found that S3 used a fraction that was not in the simplest form. He asked S10 why using "6/12" instead of "1/2". To his great surprise, S3 said it was not a mistake. "*I wanted my friend to split the piece into three parts, when they divide by "3" and use "1/2", the fraction is not friendly. I do not want my friends to have a headache.*" It was interesting to note that aboriginal students were friendly to peers and wanted the posed problem to be accepted by good friends. They did check the fractions and made sure that friends could find a solution. "*It was bad if friends cannot solve just because of my fault, the fractions I use must be ok for the problem to be solved.*"

4.2.2 Cognitive Behaviors (from Stream's Study)

Mr. Stream studied cognitive behaviors from errors. He and Susan met and found that the errors suggested students' beliefs. The result of errors was used in his subsequent teaching.

1. Decimal point not aligned

   One student believed that he could align numbers as if they were whole numbers. He ignored the decimal point and align numbers to the right.

   2-s-14. Red Fire Ants carried cookies for winter. The first time they moved 2.325kg, and the second time they moved 1.65kg. How many kilos did they move in total?

   $$\begin{array}{r} 2.325 \\ +1.65 \\ \hline 2.490 \end{array}$$

2. Carrying and borrowing error

   This student confused 1.3 with 0.13. In addition, he copied "+" instead of "−" and so, there were altogether two misconceptions in this problem.

   8-p-14. Conan originally had 3.6 boxes of cookies and bought 2.7 boxes at the market. In the afternoon, 3 friends came and ate 4.5 boxes. How many boxes of cookies are left?

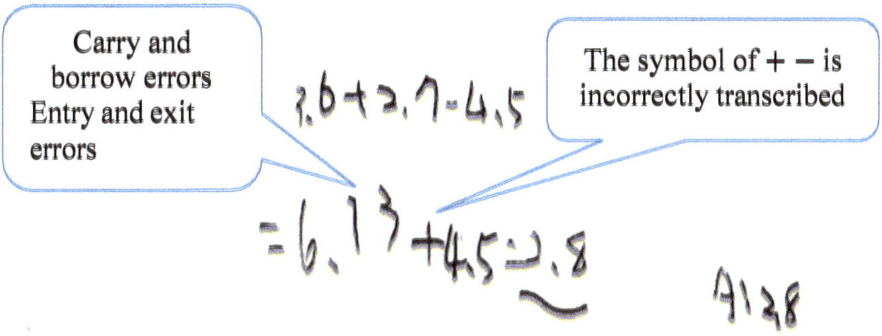

3. Calculation error in addition and subtraction of decimal

   3-p-23. The little penguin liked to eat fish. There was a bag of 5.78 kg of fish. The little penguin ate 1.58 kg, and his good friend Winter gave 3.8 kg more. How many kilos of fish were there finally?

$$\begin{array}{r} 5.78 \\ -1.58 \\ \hline 7.20 \end{array} \qquad \begin{array}{r} 7.2 \\ +3.8 \\ \hline 11.0 \end{array}$$

   After posing, the student solved this two-step problem. The carrying out was by vertical computation: first subtract (left) then add (right). On the left side, however, this student did not subtract properly: Add "5.7" to "1.5" to get "7.2" and minus 0.08 from 0.08 to get "0". This was incorrect.

4. Numeric transcription errors

   8-p-5. Conan originally had 3.6 boxes of cookies and bought 2.7 boxes at the market. In the afternoon, 3 friends came and ate 4.5 boxes. How many boxes of cookies are left?

$$\begin{array}{r} 3.6 \\ +2.7 \\ \hline 6.3 \end{array} \qquad \begin{array}{r} 6.9 \\ -4.5 \\ \hline 2.4 \end{array}$$

   From the left side of the working, the "6.3" was incorrectly copied to the right for step two and so, the answer was wrong.

Susan's note for discussion

The two teachers reported both affective and cognitive aspects of students through math problem posing. Most of his students expressed a preference for MPP. Through their diaries, Mr. Please gained insight into their heartfelt issues both at school and at home. Mr. Stream discovered that checking students' solutions helped identify and clarify misconceptions on the cognitive level.

4.3 Suggestion on How Students Learn Fractions and Decimals Using MPP

Both Please and Stream learned that students needed at least three times to practice before they got familiar with problem posing. During the pilot study, students did not know what to do in problem posing. They were frustrated and so were Mr. Please and Mr. Stream. Besides, timing is very important. They gave students enough time to think otherwise the problems were almost the same as what they come across before. They realized that for the low-motivated students, they tried to use familiar real-life situations to present materials. Mr. Please paired them with high-ability students allowing opportunities for exchanges and caring. He was excited to see students looking forward to have math lessons after using MPP.

They spent time writing notes about the activities. Mr. Please found that when students were so eager to learn more, they also made him think more about his way of teaching. About his change in teaching, he needed to use praise to motivate students. Also, to attend to cultural differences of students and watch interaction among students. As per the plan, the usual lessons were conducted. Both teachers read the diaries daily to better understand the students' perspectives. Both Please and Stream maintained journals to document their teaching methods and the student's learning processes. During regular meetings with Susan, they discussed the analysis of problem-solving worksheets, data on students' emotions, and reflections on their learning. The students' test scores showed an improvement in fractions and decimal tests after being taught through MPP. All investigators were overjoyed.

5 Conclusion

The transition of student's learning from whole number to fractions and decimals was important that deserved attention from research and practice (Behr et al., 1992; Resnick et al., 1989). The attempts of Mr. Please and Mr. Stream provided a lens to understand how problem posing can be employed in the teaching and learning of rational numbers (fractions, decimals). The part on operations involving two fractions with different denominators or two decimals with different number of places after decimals was studied. The investigators used Pose-Solve-Pose approach and students started to consider types of problems when they pose problems. First, the curriculum issue was on how fractions stood in rational number learning as it stemmed from whole numbers and later connected to decimals (Behr et al., 1983). All in all, students who did problem posing learned more about connecting math ideas. Although not yet doing problem solving, when they posed a problem, students rehearsed the solution processes and attended more to the numerators and denominators; or, checked more on the places after decimals and if the operations needed carrying and borrowing. Mr. Please and Mr. Stream found that in solving (not posing) students were not being so sensitive to the numbers. The importance of having the same units (unit fractions like halves or places after decimals like tenths) when operating two rational numbers (MOE, 2003). Although the two teachers gave reminders on checking unit fractions, or, same or different places after decimals it was not until students were doing problem posing then they remembered to do this checking. The key to this activity was, students had a heart for friends and they did not want the problem to be problematic nor could not be solved. As Resnick et al. (1989) noted, errors were hard to avoid as students made sense of new learning using what they remembered from the already learned. Thus, the teaching of rational numbers is challenging in that errors or misconceptions could be inferred from solutions or posed problems. Teachers who made good use of artifacts of lessons still could teach math topics like rational numbers.

One caveats here is to mention that the kind of MPP used here is posing a similar problem to the teacher's problems, which is one target problem from the list of problems sequencing for a unit in the textbook following the curriculum guide. In this sense, MPP is in fact a way of teaching and learning math. This time, MPP is not creating problems of a different type requested by the teachers. The journey of Please also showed that MPP could also attend to cultural differences. Textbook problems usually included the context of most Taiwanese children but in MPP, students solved problem in contexts familiar to children who were Aborigines. The fact that there was a change is curriculum standards or slight change in order of presenting materials in the case of Stream was not a problem. In MPP, teacher and students cope with chance and got used to the climate of transition of two curriculum sequence naturally using MPP. In future, more attempts on using MPP instructions can be given to math contents like ratio and proportions.

Appendix: Addition/Subtraction of Two Fractions with Different Denominators

Fraction concept		Textbook type	Examples of teachers' situational topics
Addition/ subtraction of two fractions with different denominators *denominator*	2–1 addition	1. Two fractions with one denominator as multiple of the other	Dad bought $\frac{2}{5}$ kilograms of pork and $\frac{3}{20}$ kilograms of beef to cook a big dinner. How many kilograms of meat did Dad buy in total?
		2. Fractions with two denominators co-prime	The carpenter was sawing a bamboo pole into two pieces, the lengths were then $\frac{2}{5}$ meter and $\frac{11}{12}$ meter respectively. How long was this bamboo pole originally?

(continued)

(continued)

Fraction concept		Textbook type	Examples of teachers' situational topics
		3. Mixed fractions with two denominator co-prime	Zhixiong took part in a sports competition and spent $1\frac{3}{4}$ hours cycling and $2\frac{2}{5}$ hours running How many hours did he spend in total?
	2–2 subtraction	4. Two fractions with one denominator as multiple of the other	When Uncle Wang works $\frac{8}{11}$ hours every day, when Uncle Li works $\frac{7}{22}$ hours every day, what is the difference in their working hours?
		5. Fractions with two denominators co-prime	Mother boiled a pot of soy milk, the total was $\frac{6}{7}$ liters, and my brother drank $\frac{7}{12}$ liters, how many liters of soy milk were left?
		6. Mixed fractions with two denominators co-prime	The elder brother ate $1\frac{4}{7}$ packs of biscuits, and, my elder sister ate $2\frac{3}{4}$ packs. How many more packs of biscuits did the elder sister eat than the elder brother?
2–3 Comparing sizes of two fractions with different denominators and other operations		7. Two fractions with denominators *denominator* not co-prime	Xingui's family has two flower beds of the same size, $\frac{7}{9}$ for lilies and $\frac{5}{6}$ for roselles, which one is bigger? How much in all?
		8. Two fractions with denominators co-prime	There are two pizzas of the same size, the Betty ate $\frac{6}{7}$ piece, and Sam ate $\frac{3}{4}$ pieces, who ate a larger piece? How much in total?

References

Behr, M. J., Harel, G., Post, T. R., & Lesh, R. (1992). Rational number, ratio, and proportion. In D. Grouws (Ed.), *Handbook of research on mathematics teaching and learning* (pp. 296–333). Macmillan.

Behr, M., Lesh, R., Post, T., & Silver, E. (1983). Rational number concepts. In R. Lesh & M. Landau (Eds.), *Acquisition of mathematics concepts and processes* (pp. 91–125). Academic Press.

Behr, M. J., Wachsmuth, I., Post, T. R., & Lesh, R. (1984). Order and equivalence of rational number: A clinical teaching experiment. *Journal for Research in Mathematics Education, 15*(5), 323–341.

Cramer, K. A., Post, T. R., & delMas, R. C. (2002). Initial fraction learning by fourth- and fifth-grade students: A comparison of the effects of using commercial curricula with the effects of using

the rational number project curriculum. *Journal for Research in Mathematics Education, 33*(2), 111–144.
Frobisher, L., Monaghan, J., Orton, A., Orton, J., Roper, T., & Threlfall, J. (2002). *Learning to teach number: A handbook for students and teachers in the elementary school.* Stanley Thrones Ltd.
Leung, S. S., & Silver, E. A. (1997). The role of task format, mathematics knowledge, and creative thinking on the arithmetic problem posing of prospective elementary school teachers. *Mathematics Education Research Journal, 9*(1), 5–24.
Ministry of Education (MOE). (2003). *The MOE deep cultivation project for nine-year curriculum and instruction.* http://edu.law.moe.gov.tw/LawContentDetails.aspx?id=FL029037&KeyWordHL=. (In Chinese)
Ministry of Education of Taiwan. (2008). *General Guidelines of Grades 1–9 Curriculum for Elementary and Junior High School Education* (in Chinese).
Resnick, L. B., Nesher, P., Leonard, F., Magone, M., Omanson, S., & Peled, I. (1989). Conceptual bases of arithmetic errors: The case of decimal fractions. *Journal for Research in Mathematics Education, 20*(1), 8–27.
Vergnaud, G. (1979). The acquisition of arithmetic concepts. *Educational Studies in Mathematics, 10*, 263–274.

Chapter 5
Two Studies on Learning Ratio and Proportion Using Problem Posing: Transition from Grade Six to Grade Seven

1 Introduction

In elementary school, number constitutes the major part of the math curriculum. Students' learning sequence starts from whole numbers and then to rational numbers. At the same time, students learn about the mensuration of different quantities, from familiar (lengths, weights) to novel (angles, capacity). According to Taiwan curriculum standards (MOE, 2003), students learn numbers and measures through real-life experience, and four formal operations of numbers are used to record the results of solving problems involving situations such as part-whole, taking away, and splitting a portion. Towards the upper-grade levels in elementary school and the early years of middle school, ratio and proportion appear as a challenging topic to students and teachers. Even adults are having problems in solving problems involving ratios.

In the learning ratio and proportion, the difficult part is how to "see" them in context. In learning this topic, students need to compare two quantities and associate them with a relationship. The quantities can be concrete like weights, lengths, and angular measures, or abstract such as time or speed. Hence, this concept is not easy to understand at all. Quite similar to the learning of whole numbers and operations, just a definition of ratio learning will not help. In textbooks, the introduction of ratios may start with examples and not definitions. Students experience them in contexts. To present ratio and proportion in real-life situations, the kind of contexts that appear to be meaningful and interesting to students is a question worthy of investigation. In school, the order of learning is ratio and proportion in the elementary school (grade 6), and, direct proportion, inverse proportion, and continued proportion in the middle school. The transition of math learning on this topic from elementary to middle requires revision on ratios. In addition, it also challenges the prior learning of numbers and measures in various quantities of students. Indeed, the importance is well acknowledged in research and practice (Behr et al., 1992). If students are asked to do math problem posing (MPP), the investigators know how students see ratios and proportions in context.

S. S. Leung, *Mathematics Problem Posing in Action*,
https://doi.org/10.1007/978-981-96-8480-9_5

In this chapter, two teachers decided to use MPP to study how students learn this topic. A teacher from elementary school and one from middle school. Mr. Firm taught grade 6 while Mr. Sun taught grade 7. They worked with Susan, a math teacher educator who taught at a university where they attended graduate school.

In case one, Mr. Firm designed materials with attention to semantic structure and number choices. In his review, grade 6 students learn ratios and solve problems involving two ratios, A: B = C: X, given in different semantic structures by Lamon (1993); with consideration of numbers for B and C as multiples of A or not. "What problems will students pose if I give them varying types of contexts embedded in different semantic structures and number choices?" He aimed at students' solutions to ratio and proportions problems, contexts in their posed problems, popular strategies, and common mistakes. In case two, Mr. Sun made an investigation in grade 7 by following Mr. Firm's investigation footprints in grade 6. Mr. Sun believed that math connections can be done by problem posing. "Learning ratio cannot be done in one grade. It is a long journey". He read van den Heuvel-Panhuizen et al. (1995) and followed the same design: asking students to pose one easy and one difficult problem. "When students pose, I know what problems are easy or hard for them."

The three investigators worked together. The research questions were:

1. What math knowledge did students exhibit from solving, posing, and solving problems in this unit (cognitive)? What were students' acceptance of MPP (affective)?
2. Was there any difference(s) when problems were in different semantic structures or number choices (grade 6)? And, when students were of varying ability (high, low) and posed easy/hard problems (grade 7)?
3. What did the teachers suggest after carrying out the investigations?

2 Background Literature for Action

Both Firm and Sun read literature about math problem posing. Susan shared background literature with Firm and Sun literature on this topic (e.g. Behr et al., 1992; Lamon, 1993; van den Heuvel-Panhuizen et al., 1995).

2.1 Rate, ratio, and proportion

I read Lamon (1993) *and I had never imagined how a ratio could be expressed in these wonderful contexts. Let me try to digest the content. I will learn the concept again."*

Type 1: Associated Sets

It is a ratio that represents a relationship between two elements that may have no commonly known connection (e.g., number of children and apples).

Type 2: Part-Part=Whole

A part-part-whole context involves the extensive measure (cardinality) of a single subset of a whole given in terms of the cardinalities of two or more sub-subsets (e.g., x agree to y disagree).

Type 3: Well-Chunked Measures

This type involves the comparison of two extensive measures, resulting in an intensive measure or rate (e.g., miles per hour).

Type 4: Stretchers and Shrinkers

The situation involves scaling up or scaling down (stretching or shrinking). In between two quantities, there is a one-to-one continuous ratio-preserving mapping (e.g., length and width of a rectangle).

Mr. Sun was also excited to see there were four types. *"Wow, four semantic types, I guess stretching or shrinking is most difficult..."* Later, both Firm and Sun tried to link semantic types (Type 1 to Type 4) identified by Lamon (1993) to ratio problems given in textbooks.

2.2 Rule of Three and Number Choice

Susan also shared "rule-of-three" principles during the meetings with Firm and Sun and suggested them to about multiplicative structure (Vergnaud, 1983, 1988). In a problem with ratios "1, f (1)", and "x, f(x)" in a proportion, and when three of them are known then we can use the proportion to find the fourth. Vergnaud identified types of division: Type I for finding "f (1)", and type II for finding "x". For the extension to a general case: a: b, c: x then the computation for the unknown "x" is made complicated depending on the values of a, b, and c. If one of them is "1" then it is reduced to a simple multiplication or division. Susan also reminded the two teachers to refer to curriculum standards and textbooks. The three investigators then studied the number choices given and Susan remarked, "*Number Choice is important.*"

Mr. Firm read grade six textbooks. In the curriculum guide, he counted that in a ratio and proportion problem, three numbers were appearing in the problems. The three were used to find the unknown (X). He found four cases on number choice (I, II, III, IV) assigned to the three numbers (A, B, and C).

"*These numbers might affect how my students solve problems.*" Consider four cases when given the same A but different B and C. The problem is to find X.

A: B=C:X

I. 3:6=9:X Both B and C are multiples of A,
II. 3:6=5:X Only B is multiple of A
III. 3:7=9:X Only C is multiple of A
IV. 3:7=5:X Both B and C are NOT multiple of A

He learned that Case I was the easiest and Case IV was the most difficult.

When three of the four in a proportion (A: B=C:D) are known then the student can find the fourth. As above, the unknown "X" is "D". The teachers remarked happily, "*Let us try MPP. Students can follow this "rule-of-three" and choose which of the four to be the unknown.*"

2.3 Easy and Hard Tasks Versus High and Low Students

After Mr. Sun read the textbook, he read a journal paper "Student-Generated Problems: Easy and Difficult Problems on Percentage" by van den Heuvel-Panhuizen et al. (1995). In that study, students in grade 5 were asked to generate one easy and one difficult problem. The range of problems given by students was professional and included a variety of problems that almost covered the whole part percent of the curriculum. The reasons for easy/hard were: the size of numbers, tools, done before or not, and structure. Mr. Sun decided to see how his students posed easy and difficult problems on proportions. "I shall check if my students were also professionals and posed all kinds of proportion problems in the unit." Finally, he was impressed by the work on expert-novice differences (Chi, 1984). Chi found that children's classification outcomes are a function of the knowledge that they have and the representation that the knowledge takes. While experts classify according to domain structure, novice children classify on superficial features. "What is easy to a student might not be easy to another…" He decided to study if there was any difference in posing easy/hard problems given by students of varying abilities in mathematics.

3 Method

3.1 Participants

3.1.1 Case of Grade Six (Mr. Firm & His Students)

The teacher was Mr. Firm. Mr. Firm was the class teacher and also taught math in the class. He represented a teacher with determination. Whenever there were professional development sessions on math education, he sent in registration and showed up. His school and home were both located far from the university for graduate school but he drove for his goal. Loving his job and children made him smile every day. One day he read Lamon (1993) and he remarked, "My students often find difficulty with ratio" When it was time for him to decide on a study for his master's thesis, he tried problem posing ratio problems. Students were 30 grade six students in an elementary school in a rural region. There were 15 boys and 15 girls. This class was one of five from the same grade. The academic performance was this class was average, which meant comparable to the other four classes. This school was in the countryside. Parents were mostly workers and had their businesses. When they communicate with Mr. Firm, they would use phone calls.

3.1.2 Case of Grade Seven (Mr. Sun & His Students)

Mr. Sun was a teacher in middle school for 15 years. He was a mathematician then got teacher qualification and taught grade 7 to grade 12. To him, curriculum standards changed over time in his career but the part on ratio and proportion stayed almost the same. He taught seriously and shared his work with colleagues. As a senior teacher in the school, he could ask for help and support easily from peers and parents when in need. His students described him to be a prepared teacher and when he assigned homework students would not say, "No." Students who received instruction with math problem posing (MPP) was one grade seven class: 30 students (15 female and 15 male). Students were divided into three groups according to math performance (High, Medium, and Low). All other classes of grade 7 in the middle school adding to 120 more students were invited to make up easy/hard problems for the whole unit. Therefore, a total of 150 students participated in this study.

3.2 Design of MPP Tasks and Program

3.2.1 Design for Grade 6 (Firm: rate and Proportion)

MPP Tasks

Susan and Firm decided to use a design called solve-pose-solve, meaning, students solved a problem then posed one on their own, and finally solved their problems. Mr. Firm considered his students as diligent followers. He believed that by using a structured approach, he and his students would feel comfortable. "*I think this is the approach my students and I would like best*".

Tasks were combinations of ratio problems using Lamon's semantic types (1, 2, 3, 4) and cases on number choice (I, II, III, IV). There were altogether 20 items. First, Mr. Firm developed 5 items using four semantic types by Lamon (1993). At first, he made one item per type. Later, he found that there were more familiar contexts for children in semantic type 1 (Associated Sets) and so he made 2 items in that category.

Type 1: Associated Sets (2 items, 1–1, 1–2)
Type 2: Part-Part-Whole (1 item)
Type 3: Well-Chunked Measures (1 item)
Type 4: Stretchers and Shrinkers (1 item)

Second, he referred to the four cases and played with the numbers. In each of the above 5 semantic types he made 4 problems corresponding to each case in number choice (I, II, III, IV), resulting in a total of 20 items. Take the shadow problem as an example. *There is a tree with a height of 180 cm. The shadow length of the tree is 540 cm. If Xiao Ling's height is 150 cm, how long will Xiao Ling's shadow be?"*

The semantic type by Lamon (1993) is *Stretchers and Shrinkers* (Type 4). As for number choice, only 540 is a multiple of 180 (type II).

Firm's Teaching Program

Stage I: Problem Solving

1. (15 min) Teacher's presentation of the problem (worksheet to students)
   Students solved problems and handed worksheets back to Mr. Firm. He analyzed the worksheets and prepared for discussions (Types of solutions and mistakes)
2. (15 min) Whole class discussion: drew attention to different strategies. Another worksheet was given to students. Students made a record of what they learned from the discussion.
3. (10 min) Solutions: note-taking of methods and learning about ratios

Stage II: Posing and then Solving

1. (15 min) Students posed a problem Students solved their posed problems
2. (15 min) Sharing with peers
3. (10 min) Whole class discussion

After class, Mr. Firm completed journals while students wrote math diaries.

3.2.2 Design for Grade 7 (Sun: Proportion, Continued Proportion, Inverse Proportion)

MPP Tasks.

Susan and Sun used the same design, "Solve-Pose-Solve". Students were given a problem to solve, then they posed one problem and asked friends to solve it. In later stages, they also made easy/difficult problems and gave reasons. His designated content is proportion. This unit was an extension of "ratio and proportion", taught in grade 6.

Sun's teaching program

Stage I: Solve-Pose-Solve (Sun's class, $n = 30$)

After teaching the lesson on a topic, students completed a worksheet on that lesson. Mr. Sun presented one typical problem for that lesson, students solved then posed one, and then a friend solved it. "*Pose a problem to check how much your friend learned.*" They also needed to ask Mr. Sun, "*Why did you pose this problem to challenge your friends?*" This lesson was repeated four times: proportion, continued proportion, direct or inverse proportion, and revision. For the last lesson on revision, students posed freely any problem in the unit. Take continued proportion as an example, "*The ratio of the number of students walking and riding bicycles is 3:5, and the ratio of the number of students riding bicycles and their parents picking them up is 3:1, so what is the ratio of students walking, riding bicycles and their parents picking them up?*".

Stage II: Pose Easy/Hard problem
Students posed two problems (one easy, one hard). They checked a reason or wrote their reasons for considering the problems as easy/hard.

3.3 *Data Collection and Analysis*

3.3.1 Data Collected from the Case of Grade Six (Firm)

Mr. Firm collected data including (1) Problem Solving Worksheets, (2) Problem Posing Worksheets, (3) Math Diaries Sheets, and (4) Teacher's Journals. The worksheets were randomly assigned worksheets so that students had a balance in semantics types (1–4) and categories of number choices (I–IV).

1. Problem Solving Worksheets
   e.g. Type 1, Category I
   Dear problem solvers, please complete the following question and write down your thoughts at the same time. You can also use pictures to answer these questions. Good luck with your problem solving.
   [1] If it costs $24 to buy 6 apples, how much does it cost to buy 48 apples?
   My answer is: 【 】
2. Problem Posing Worksheets
   Students did MPP right after solving one particular problem (one type of semantics and one category of problem choice). The researchers also designed a worksheet for Problem Posing just in case they forgot to do so.
   After the last class, you became a little expert in solving problems; now, please pose a similar question based on the topic of the last class, and solve by yourself. Remember that its solution should be similar to the last problem!
   【The question I posed】
   【Please share your solution】
   A:________
3. Math Diary Sheet
   After the MPP class, students wrote math diaries. Given that students might not write diaries about the MPP lesson, Mr. Firm let students experience writing by answering three guided questions. Students could write more of their choices after these 3 questions.

   1. What do you learn after today's lesson?
   2. What do you think about the problem you posed today? If it is bad, how should you change it to make it good? If it is good, please tell me why you think it is good.
   3. What do you feel after today's math class? Please write down your thoughts.

4. Teacher's Journals
   Mr. Firm wrote journals in a free format by referring to artifacts of each lesson and talking to students after class.

Data analysis (Firm)
On problem solving, as long as the solution was reasonable, it was counted as correct. There was no restriction on the mode of explanation or strategy. Below is one correct solution. [Type 1, category I] *If it costs $24 to buy 6 apples, how much does it cost to buy 48 apples?* $48 \div 6 = 8$, $24 \times 8 = 192$.

My answer is: 【 192 】
After the Firm collected data, he and Susan did an analysis quantitatively and qualitatively. Each response was coded according to Leung and Silver (1997) and then considered the semantics and number choice.

3.3.2 Data Collected from Case of Grade Seven (Sun)

Data sources were: Solve-Pose-Solve worksheets (n = 30, Stage I), Easy/Hard problems posing worksheet (n = 30, Stage II). Individual interviews (4 high and 4 low students), and, Teacher's journal (after every lesson). Mr. Sun wrote a journal after every lesson, to check individual differences and also how he attempted to find solutions to teaching problems arising over time.

As above, the two teachers collected students' worksheets (n = 30) and teachers' journals. While Firm collected student diaries Sun conducted interviews with eight students.

4 Results

4.1 Math Knowledge Exhibited (Cognitive) and Students' Acceptance (Affective)

4.1.1 Case of Grade Six (Mr. Firm): Solve-Pose-Solve

What Math Knowledge Was Exhibited (Cognitive)?

Stage I: Solving the teacher's problem
The teacher did not just aim at problem-posing. He invited students to solve the problems. Out of 600 solutions, 465 were correct, and, 135 were incorrect.

The 465 correct solutions were further analyzed. Strategies for solving them correctly were four: Multiple (n = 268), Unit Price (n = 174), Formula (n = 12), and Additive (n = 11). The following were descriptions and examples, in descending order of the number of occurrences.

1. Multiple method (44.7%): Find out how many times as much as the first purchase, then multiply the amount for the first purchase. "The price for 6 apples was \$24, how much will 48 apples cost?" $48 \div 6 = 8$;$8 \times 24 = 192$ (3-s-1)

   This student noticed 48 apples is 8 times as much as 6 and therefore used this strategy.
2. Unit price method (29.0%): Find how much each item cost, then multiply the number of items.

   "The price for 6 apples was \$24, how much will 48 apples cost?" $24 \div 6 = 4$;$4 \times 48 = 192$ (1-s-6)

   This student found the price of an apple then times 48. It is a very effective way of solving problems concerning purchases and when involving whole numbers.
3. Formula Method (2%): Equate the two ratios to get an equation and then cross multiply and get the answer.

   The price for 6 apples was \$24, how much will 48 apples cost?" (4-s-19)

   6: 24 = 48: □

   □ = $24 \times 48 \div 6 = 192$ (4-s-9)

   This strategy only appeared for problems with number choice IV (Both B and C, not multiple of A) and so whole number Multiple and Unit price methods are not easy to use. For the formula method, the students use equations and find the unknown. For grade 6 students, they usually used □ to represent the price of 48 apples. The problem was a two-step problem.
4. Additive Method (1.8%): Adding \$24 for every 6 more apples, finally get \$192.

   The price for 6 apples was \$24, how much will 48 apples cost?" 6 ➜ \$24; 12 ➜ \$48; 18 ➜ \$72; 24 ➜ \$96; 30 ➜ \$120; 36 ➜ \$144;

   42 ➜ \$168; 48 ➜ \$192

   Again, this strategy only appeared for problems with number choice IV (Both B and C, not multiple of A) so whole number Multiple and Unit price methods are not easy to use. The student used \$24 as a price for 6 apples, then for an addition of 6 apples he added another \$24. He added "24" many times until he got 48 apples, and the price was \$192.

After addressing the 465 correct responses, the 135 incorrect solutions were further analyzed. The reasons were five: random calculation (n = 59), mistakes in the location of terms (n = 27), computation mistakes (n = 24), ignoring partial information (n = 18), and misused data (n = 7). The following were descriptions and examples, in descending order of the number of occurrences.

1. Random calculation (59/135; 9.8%)

   When students can't understand the problem, they add, subtract, multiply, and divide the numbers in the problem. For example: A and B are two similar rectangles (as shown in the diagram). The length of A is 9 cm and the width is 5 cm; the length of B is 16 cm. What is the width of B? 【Question 17】 The student's solution record: "16–9 + 5 = 12 (4-s-13)". "16–9 + 5" is meaningless.

2. Mistakes in location of terms (27/135; 4.5%)

   Students located the terms in a reverse position in a proportion and failed to solve the problem. For example, "The price of a ribbon is determined by its length. If we used \$12 to buy a ribbon that is 8 cm long, how many centimeters of ribbon can we buy for \$20? 【Question 18】 " Students' problem-solving records: (4-s-14) \$20 × 12 ÷ 8 = 30. The student originally wanted to use this proportional relationship to find the answer, but reversed the ratio "12 ÷ 8", resulting in an error.
3. Computation Mistake (24/135; 4.0%)

   Careless mistakes. For example, "Grandpa has 100 acres of land, and the ratio of the total area to the area planted with fruit is 12:7. What are the acres of the area planted with fruit? 【Question 20】 " Students' problem-solving records: (4-s-19) $\frac{100}{1} \times \frac{12}{7}$ should be $\frac{175}{3}$. The student multiplied the wrong number and failed. (The correct answer should be 58)
4. Ignore partial information (18/135; 3.0%)

   Many children often ignore important information or data which makes mistakes in solving problems (Hart, 1981; Karplus et al., 1983). For example, two equilateral triangles have sides with a ratio of 9:6, and the perimeter of the large triangle is 45cm. What is the perimeter of the small triangle? 【Question 13】

   "Students' problem-solving records": (3-s-18)45 ÷ 9 = 5, 6 × 5 = 30 but 30 is the side, not the perimeter.
5. Misused data (7/135; 1.2%)

   This was caused by students' misunderstanding the meaning of the problem or not understanding the meaning of ratio. For example, "*Grandpa has 100 acres of land, and the ratio of the total area to the area planted with fruit is 12:7. How many acres of the land were planted* with fruit?" 【Question 20】 "Students' problem-solving records: (4-s-29)The student added "12" to "7" and "19" represented the total number of acres. Actually, "12" represented the whole and not the part. The part-whole relationship was misused.

Stage II: Posing-then-solving

Students posed then solved and shared with peers. Mr. firm conducted discussions.

Difference between the posed problems and the teacher's problem About the ratio problems students posed, over 95% of the problems were solvable math problems. Students used the structure of ratios and assigned proper numbers for the ratio problems. The problems were similar to the ratio problem presented by Mr. Firm during Stage 1. A total of 117 problems were posed.

1. Changed the number (117/117).
2. Changed the objects (85/117; e.g. "Apples" was changed to "eggs".)
3. Changed the structure of the problem (30/117).

For example, the original problem was to find distance given speed and time but the new problem was to find time given speed and distance. (3-p-13) *My sister can drive 150m per 10 s. It is 90km from Qishan to Kaohsiung. How many hours will it take her from Qishan to Kaohsiung?*

Knowledge of students exhibited (Cognitive side 2)
Mr. Firm gave instructions in the worksheet:

> *"After the last class, you became an expert in solving problems; now, please pose a similar question based on the topic of the last class, and solve by yourself."*

Rule of Three. Among students' responses, most problems were posed by using the principles of rule-of-three. Below is one typical example (from the third lesson S3).

【*The question I posed*】

There are two regular hexagons, and the perimeter ratio is 42:36. The perimeter of the big regular hexagon: ① If it is 126cm, how many mm is the length of the side of the small regular hexagon? ② If it is 504cm, what is the perimeter of the small regular hexagon in cm?

【*Please share your solution*】

① $126 \div 42 = 3$, $3 \times 36 = 108$, $108 \div 6 = 18$, $18 \times 10 = 180$

A:(180) mm.

② $504 \div 42 = 12$, $12 \times 36 = 432$ A:(432) cm

Side and perimeter in proportion. The student who posed this problem knew that the perimeters of two regular polygons (big, and small) were in the same ratio as the sides (which is one-sixth of the perimeter). The students did not forget to convert the measure of the side from a larger unit to a smaller one (cm to mm). Both problems were solved by attending to multiplicative structures (Vergnaud, 1988). Note that the students were attending to good number choices and that 126 and 504 were multiples of 42. The numbers thus make computations friendly.

Patience and alertness (Affective side)

Solving patiently. The other students did not provide good numbers and Type IV was posed. That is, both are not multiples. Yet, the student solved patiently. His friend asked him "If the numbers were not given by you, will you solve it patiently?"

<u>The fourth lesson 6-A (S13)</u>

【*The question I posed*】

Xiao Ming is going to school to play, he can walk 30m every 43 s, and it takes 10 min to walk to school, how far is the school?

【*Please share your solution*】

Table 1 Feedback from math diaries

	Reward from MPP lessons				Understanding			
	A lot	Quite a lot	A bit	No	Full	Mostly	A bit	No
First	19	9	2	0	4	23	3	0
Second	12	13	5	0	12	13	5	0
Third	14	15	1	0	10	16	4	0
Fourth	16	14	0	0	9	20	1	0
Total	61	51	8	0	35	72	13	0
Percentage	50.8%	42.5%	6.7%	0.0%	29.2%	60.0%	10.8%	0.0%

$10 \times 60 = 600, 600 \div 43 = \frac{600}{43}$,

$\frac{600}{43} \times 30 = \frac{18000}{43} = 416\frac{12}{43}$

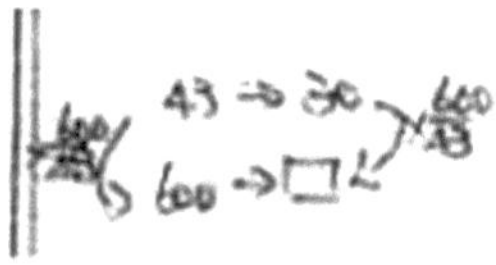

Responses to problematic problems. When students posed but did not solve, they did not realize if the problem was possible to solve. One student wrote a problem on finding the area of a square with a side 4 cm and an area of 10 sq. cm. The students did not notice that the problem was implausible to solve. According to Mr. Firm's program, they did problem posing and solving their own problems during Stage II. During Stage II, the student learned that the particular problem was problematic.

Students' Acceptance (Affective)

From Table 1 and math dairies.

As in Table 1, students were satisfied with their performance in posing and solving problems. The amount of responses in "A Lot" of rewards is half. For Understanding, 60% checked "mostly" and about 30% checked "Ful" understanding. Also, there was an increase in percentage over time (from First to Fourth).

Math Diary Sheet.

a. What did you learn after today's lesson?

   Students realized that they could use drawings or diagrams to represent proportion problems. You can use drawing to analyze it [the problem]. It will be easier to solve because drawing is relatively simple. (S3) I learned how to draw diagrams to make it easier. (S17)

   They compared it with other topics and considered this topic more challenging after doing problem-posing. *Learning ratio and proportion is not easy. (S23) Compared with the last time, I think the topic this time is now more challenging (S30)*

b. What do you think about the problem you posed today? If it is bad, how should you change it to make it good? If it is good, please tell me why you think it is good.

They also used "good' or "bad" (given in the question sheet). When they said "good" they might or might not give a reason. *I think it is good. (S3) Good, I know that 63:9 can be used, divisible! (S1) I use good numbers so that friends can complete the calculations in the shortest time. (S10)*

However, when they said "bad", they tended to give a reason such as similar or too simple. *Not good. Most of the questions posed today are based on the exam papers (S5) The questions posed are too simple. I will make it more difficult in the future. (S17)*

c. What do you feel after today's math class?

Feelings such as freedom, thinking, know-how, and attitude toward math.

I LIKE it, I can freely express my own ideas. (S3)

I will think about it carefully before writing the topic (S30). It will take me quite some time to think about it. (S10)

As long as I know the ratio, it will be easy to pose the topic. (S17) Problem posing completely changed my original idea of studying by memorizing mathematics, and now I can learn mathematics by understanding. (S1)

4.1.2 Case of Grade Seven (Mr. Sun): Solve-Pose-Solve

What Math Knowledge Was Exhibited (Cognitive)?

Problems from each topic were posed: proportion, continued proportion, and inverse proportion. Mr. Sun also found outcomes on students' knowledge in ratio, which reminded him of the findings of Mr. Firm in grade six. Students posed problems in all four semantic categories by Lamon (1993). He added below what he found but not yet reported by Mr. Firm. Sun analyzed them according to the order in the textbook of grade seven.

Proportion

There is a distance of 1000 m. It takes A 25 min and B 1.5 h to reach.

What is the ratio of the average speed of A to the average speed of B?

Continued proportion:

There are A, B, C three pieces of rice blood, weight ratio 2: x: 9, thickness ratio y: 3: 11, B's thickness = 2, A's weight = B's thickness, xy = 18.

Direct/inverse proportion:

Xiao Ming's running time is inversely proportional to the distance he runs per second.

When running 10m per second, it takes 10 s to get from A to B.

When he takes 1 s, how many meters per second does he run?

When they reviewed friends' problems, they were professional. Below are two problematic problems and friends' feedback. The first is a problem with multiple choices but none of the choices was correct; and, the second is a problem without a question.

A problem with no correct options

Two jars are filled with the same weight of alcohol, and the weight ratio of water to alcohol is 4:5 and 3:5 respectively. If the two jars of solution are poured into a larger container without overflowing, then what is the weight ratio of water to alcohol in the result? (A) 4:10 (B) 7:5 (C) 7:10 (D) 12:5.

Mr. Sun read and saw that none of the four was correct. Before Mr. Sun gave any response, the friends attempted this problem and gave feedback, There was no solution. Please add 59:85.

A problem with no question

A, B, and C are three pieces of cheese, the ratio of weights is 2:x:9, the thickness ratio is y:3:11, the thickness of B = 2, the weight of A = thickness of B, and xy = 18.

When Mr. Sun was about to suggest adding a question he heard from the floor, "Where is the question? Ask and we will answer it." Therefore, he praised the students for asking "Where is the question?".

Wrong category

There were examples of posed problems given in the wrong category such as direct proportion as inverse proportion. Mr. Sun would follow and check if this was an indication of math knowledge (concepts on direct and inverse) or just being careless about language and mixing up the two in terminology.

Improvement over time

The activity, MPP, required math knowledge, and even when students did review and sent feedback to peers. By the end of the unit, all students improved in posing problems, even Low-group students posed legitimate problems, and one special student (with a language problem) also posed proper problems. Mr. Sun compared the math performance with the previous semester and found that the whole class improved in math performance over time.

Students' Acceptance (Affective)

1. *Do you LIKE math problem posing?*

   For those who responded with a liking in writing, they wrote to Mr. Sun the following:

   They expressed a liking for their own learning such as, "I can try to solve others' easy/hard problems", "It's time to check if my brain is gone" or for fun, and "I can trick my friend by posing strange problems". Some behaved as a mature teacher, "some posed messy problems, if all of them posed seriously then learning proportion would be fun and rewarding." They also did MPP as reflection, "Yes and no on liking, some problems were clearer when I solve them again." It means that students remember problems they solved before. As Mr.

Sun promised to include good posed problems in the test those who remembered problems were confident, "We now know approximately the direction of test items by Mr. Sun" or "Yes, sounds brilliant but not so…ha…"

2. *Can you solve the problem posed by yourself?*

    I can solve the problem I posed but if you asked me if my friends can I would think it depends on who.

    MPP is challenging. You need to understand concepts before posing a problem. When solving the problem, you also review the problem type. The teacher does not need to keep talking about this topic.

    I was super attentive in class; it would be better to collate problems given by everyone into a test paper. Mr. Sun, please continue to teach MPP next semester.

4.2 Differences in MPP Outcomes

4.2.1 Differences Due to Semantic Structure and Number Choice

Mr. Firm addressed to task variable in this grade 6 class.

For semantic types. Associated Sets and Part-Part-Whole problems appeared to be slightly easier than the other two types. Stretchers and Shrinkers were the most difficult type. Students' performance in solving the five different semantic types was highest in the Associated Sets and Well-Chunked Measures (both 88.3%), followed by the Part-Part-Whole (76.7%) and the Associated Sets (74.2%), and the lowest was the Stretchers and Shrinkers with only 60.0%. According to Lamon's (1993) study, the Stretchers and Shrinkers questions were the most difficult because students were unable to identify the meaning of the multiplicity of the problem situation. The results of this study were consistent with Lamon's (1993) study. Mr. Firm was correct in predicting the difficulty concerning semantic types.

On number choices. Type IV (when none of B nor C was multiple of A) appeared to be the most difficult. *"My preparation by reading related literature, curriculum standards, and textbooks was so important. And I am glad I chose the type of posing task, which was considerably friendly to my dear students."* Hence, Mr. Firm was also correct in predicting the difficulty concerning number choices.

4.2.2 Differences Due to High Versus Low Students

Mr. Sun addressed a subject variable in this grade 7 class (high, low).

Mr. Sun reported on students' completion of MPP. The posed problems showed students' structure of connected knowledge pieces versus single pieces in isolation. *"My dear students re-grouped information in the tasks when they did MPP. How nice? As I read, I realized they managed to use a posed problem to represent the structure they built."*

Students from high groups tended to pose problems with connections to other knowledge pieces or included complicated numbers.

Students from low groups tended to pose problems that referred to single knowledge pieces.

"This multi-versus-single concept by high/low student is really obvious."

Mr. Sun flipped the MPP worksheets and shared with excitement his findings with his colleagues.

Reasons for easy/hard
He interviewed students and some interesting responses were noted. High students commented on ratio and proportion structure *"When the area of all rectangles is 24 sq. cm then the length and breadth are in inverse proportion."* and experiences in problem solving *"A problem we have not come across before"*. Low students were affected by the context of the problem or prior actions of students and the teacher. *"Because many of us could not understand it."* Or, *"Mr. Sun told us already, then, it should not be hard."* The reason could be due to the numbers, *"The three numbers, 5, 10, 15 are easy to compute."*

High versus low cognitive demand
In Stein et al. (2010), there were four types of tasks: tasks by memorization, tasks without connections, tasks with connections, tasks for doing math. These four types of tasks ranged from low to high cognitive demand. Mr. Sun found reading posed problems from MPP lessons. He found that students from high groups tended to pose problems with connections or problems for doing math. *"They posed proportion problems that used concepts from prior units such as H.C.F & L.C.M."* Students from low group tended to pose problems without connections or could be completed by memorization. *"In a class of 20 students, only 8 were girls. Find the ratio of girls in this class."*

In all, the easy/hard problems from the two groups were different in the above aspects. Besides, there were expressions from low students like below, which were not related to reasons for easy/hard problems: "*I do not know*", "*Pressed with time*", or, "*My friends were very clever*".

How high/low students considered posing Easy/Hard problems About hard problems, the reasons were many and it could be a hard topic. Some students considered topics as the reason. One student exclaimed, *"All solvent problems were hard problems."* Another reason for hard was when there was a hidden required step involving a knowledge piece. For example, the part on continued proportion was required in the problem, "a:b = 2:3 and b:c = 4:5, a + b + c = 70, find a, b, and c." However, the problem did not ask students to find a: b: c. If the problem was changed to, "a: b = 2:3 and b:c = 4:5, a + b + c = 70, find i) a: b: c, and ii) find a, b, and c." hinting an intermediate step, then the problem was not so difficult.

Also, only students from high groups posed problems requiring correspondence. These students sent in problems with carefully planned steps way ahead. Mr. Sun found that the knowledge structure presented by his students in their work was consistent with the results of Chi (1984). Students in high groups organized and extracted the relevant mathematical knowledge they had learned, and used this mathematical knowledge to formulate a problem. They recalled related chunks from their memory or bank of prior knowledge (such as factors, multiples, simultaneous solutions, and graphs of simultaneous equations).

4.3 Suggestions from Firm and Sun

After teaching by using MPP, what did the two teachers suggest? Both Mr. Firm and Mr. Sun found that the ideas might not be restricted to problem posing, although the lessons he reflected upon were from the MPP program. In the flowing, the "Italics" were taken from teachers' journals and students' diaries or interviews.

1. Decision on instructional materials

   "*In doing Solve-pose-solve our dear students re-considered the original problem and in posing they revised the structure of the problem.*" Firm's choice of MPP task, semantics structure for ratio, and sets of three numbers was appropriate. Students realized that the numbers they put into the problem were important. In all, he was satisfied with his design of the MPP program for teaching ratio and proportion. Sun reminded teachers to use integration and not separate posing from other math lessons; no extra time was required. He agreed to Firm's design on Solve-pose-solve. He would try example problems to check the appropriateness: amount, order of presentation, difficulty level, and time use. In trying MPP, the teacher must know the topic well. He or she then considered the nature of the content and chose an appropriate type of MPP task. Both teachers consider math topics, like ratio and proportions, with connectedness to be great for problem posing.
2. Belief and determination

   Ratio is hard. Students were followers and inexperienced in problem posing. Mr. Firm believes and remembers the advice from Susan. "*There wasn't any right nor wrong material: just think of what you think is best to teach and for whom.*" Proportion is not easy. Mr. Sun believed that students with different abilities benefited from problem posing. For example, in this attempt, they were finding relevant mathematical knowledge to and from posed mathematical problems. Even when things did not go well or as expected, Sun gave himself a reminder, "*When I was in doubt, I just tell myself, MPP is not a panacea, I can try another teaching strategy. Do not get frustrated if it does not work.*"

   They were determined to do preparation and follow-up work. When students solve, no matter correct or not, the teacher must check. After class work, the teacher study students' work, and carefully selected work to be shared and used in teaching. "*Do not randomly choose student work to avoid the low quality of discussion"*. When the work of students suggests misconceptions, the teacher shall respond with individual interviews and remedial teaching.

3. Attending to discussion

   Discussion culture is a condition. Mr. Firm gave credits to participation. *Shall we give credits to those who reply, also to those who ask?"* When getting praise, students behaved better than before. *"We like this question, shall clap our hands?"* When quiet, Mr. Firm pretended that he heard someone trying to say something like, "I heard that someone wanted to reply, is that you?"

   To encourage voicing, he might ask students to guess what their friend meant, *"he wrote, $24 \div 8 = 3$, why did your friend wrote this expression as solution, can you try to tell us?"*

   In the grade seven class, Mr. Sun used the projector to display selected work for whole class discussion. When discussing peers' work, he found that the students in were more attentive and focused than usual. Students proposed friends as figures in the posed problems work and often attracted peers as protagonists; sometimes appeared as unreasonable or even disturbing. Such situations might be a fun time for everyone. They must pay attention to listen to problems like "Alex (a name in the class) grew 900 cm after summer holidays" and laughed their hearts out.

4. Attend to low achievers

   Mr. Firm and Mr. Sun split students into small heterogeneous groups. They asked questions and asked each group to try and final answer that the whole group agreed. *"Even if one of you disagreed your group must check again and again until all agreed"*. In this way, the students would teach each other. Although Mr. Firm pretended that he randomly picked students to answer, he tried to name low achievers as well. The low achievers, after confirming themselves in small groups, voiced boldly and became more confident when Mr. Firm "randomly" picked them to answer.

   "*More interesting than before doing MPP*" was a remark from a *high* student of Sun's class. However, he recited this remark bolding and loudly in the whole class discussion as if many students wrote the same note to him. In MPP, when a lesson appeared to be more interesting students actively participated. Finally, he happily read, "*Today"s presentation is more enthusiastic than the previous one.*" It was a note from a low student. The motivation passed from high students to *low* students. After Sun's interview, Sun wrote in his journal, "*After Ming's got 5 points for the group because of the last presentation yesterday, he raised his hand three times today*". It seems that he's willingness to learn has obviously become stronger.

5. Students responses

   It is great that this student has included the concept of unit conversion in her problem. *"Think about it again. Could you do it too? Or could you come up with even better and different problems?"*

   Students expressed in diaries that reading peers' work improved their way of posing problems. For example: "*There are 63 steps in the ladder, and Ming has walked 26 steps and still has 38 steps to go. If Ming walks 37 steps, how many steps will he have to go?*" (Problem posed by Xin Yi's). The student immediately said that it was not a ratio problem. Firm remarked, "*It shows that students understand what ratio problem is.*"

As students needed to solve after pose, students played with the "givens and goal" versus "known and unknown". They found a magic rule and were happy to conclude "*Think of two ratios or four known numbers, then take away one to make it unknown in the posed problem*" Then they thought posing a problem was so easy.

5 Conclusion

First, learning is a continuum, and "ratio and proportion" is only one example topic in need of extension from elementary to middle school. Working for a smooth transition is challenging for learners in this age group. In using MPP as a means, the end is to make students learn this topic and link it to prior knowledge learned a year before. Through students' outcomes from carefully designed MPP tasks, the two teachers (Firm: grade 6, Sun: grade 7) knew how students "see" ratio and proportions in contexts given. A linkage between grades is important as spiral curriculum ideas were adopted in Taiwan's math curriculum.

Second, variables in MPP were considered. In these two cases, the investigators witnessed the growth of students using planned activities: solve-pose-solve and systematic variation of *tasks* and *subject* variables for two different grade levels as *situation* variables (Kilpatrick, 1978). An example on task variable in Firm's case is, even with the same set of number choice, problems on ratio appeared to be more difficult for semantic structure like *shrinkers* (Lamon, 1993). For subject variables, see in Sun's case, high students posed problems by attending to structure whereas low students changed superficial features. The difference in high/low students is consistent to prior results on expert-novice difference in performing categorization tasks (Chi, 1984).

Third, the results yield important implications for learning (and teaching). The teachers did not teach yet students did learn. Note that the type of MPP here is fixed and the order of each task follows a specified sequence (Behr et al., 1992). The teachers did not teach the "rule-of-three" (Vergnaud, 1983, 1988). However, when students did MPP many times over time they learned the "rule-of-three". *Knowing three of the four in the relation you can find the fourth.* When doing MPP, students were considerate and made sure the numbers were good for solving. It was found that when MPP was in use, there was learning from students in elementary and middle school, and, had the potential to high school. In recall, the integration of MPP for varying tasks and subjects was given in the preceding chapters, from grade 2 to grade 7. Future investigations may consider a case in high school.

References

Behr, M. J., Harel, G., Post, T. R., & Lesh, R. (1992). Rational number, ratio, and proportion. In D. Grouws (Ed.), *Handbook of research on mathematics teaching and learning* (pp. 296–333). Macmillan.

Chi, M. T. H. (1984). Knowledge-derived categorization in young children. In D. R. Rogers & J. A. Sloboda (Eds.), *The acquisition of symbolic skill* (pp. 327–332). Plenum Press.

Hart, K. M. (1981). Ratio and proportion. In K. Hart (Ed.), *Children's understanding of mathematics*: 11–16 (pp. 88–101). John Murray.

Karplus, R., Pulos, S., & Stage, E. (1983). Proportional reasoning of early adolescents. In R. Lesh & M. Landau (Eds.), *Acquisition of mathematics concepts and processes*. Academic Press.

Kilpatrick, J. (1978). Variables and methodologies in research on problem solving. In *Mathematical problem solving* (pp. 7–20).

Lamon, S. J. (1993). Ratio and proportion: Connecting content and children's thinking. *Journal for Research in Mathematics Education., 24*(1), 41–61.

Leung, S. S., & Silver, E. A. (1997). The role of task format, mathematics knowledge, and creative thinking on the arithmetic problem posing of prospective elementary school teachers. *Mathematics Education Research Journal, 9*(1), 5–24.

Ministry of Education (MOE). (2003). *jiāo yù bù jiŭ nián yī guàn kè chéng yŭ jiāo xué shēn gēng jì huà* [The MOE deep cultivation project for nine-year curriculum and instruction]. http://edu.law.moe.gov.tw/LawContentDetails.aspx?id=FL029037&KeyWordHL=

Stein, M. K., Smith, M. S., Henningsen, M. A., & Silver, E. A. (2010). *Implementing standards-based mathematics instruction*. Teacher College Press.

van den Heuvel-Panhuizen, M., Middleton, J. A., & Streefland, L. (1995). Student-generated problems: Easy and difficult problems on percentage. *For the Learning of Mathematics, 15*(3), 21–27.

Vergnaud, G. (1983). Multiplicative structures. In R. A. Lesh & M. Landau (Eds.), *Acquisition of mathematics concepts and processes* (pp. 127–174). Academic Press.

Vergnaud, G. (1988). Multiplicative structures. In J. Hiebert & M. Behr (Eds.), *Number concepts and operations in the middle grades* (pp. 141–61). NCTM.

Chapter 6
The Learning of Probability for Vocational School Students Using Problem Posing for Real-Life Situations

1 Introduction

Although probability already deserved attention in both psychology and mathematics education a long time ago (Fischbein, 1975; Piaget & Inhelder, 1975), probability is still not yet considered a major area of research in mathematics education (Chernoff & Sriraman, 2014). According to Langrall (2018), most primary schools in the US already eliminated probability from school curricula. Therefore, in research and practice, the importance of probability still needs attention and recognition. The teaching and learning of probability can stretch from kindergarten to undergraduate levels (Batanero & Álvarez-Arroyo, 2023). Even young children can get an idea of randomness from colored jelly beans or have some experience in combinational reasoning through dressing a doll. On the other hand, even after attending school, adults find probability difficult and may have biases in probabilistic reasoning. Prior research suggested that visualization tools might help: tree diagrams, two-way tables, icon arrays, or Venn diagrams. Despite the aid of tools, the teacher was in doubt and faced challenges in the teaching of this topic.

Thus, there is a need for more research into teaching probability and statistics to students from different grade levels, and, for learners with different characteristics. In Taiwan, compulsory education for all students from grade one to grade twelve was adopted in the year 2014. Starting from grade ten, there are two syllabi for mathematics given in the official curriculum standards document (Syllabus A, Syllabus B; MOE, 2018). Syllabus A is for ordinary high schools and syllabus B is for vocational schools. The difference between A and B was not apparent as the names of the chapters are about the same. However, problems in syllabus A are more challenging. In learning probability, the part on Bayes Theorem is not in syllabus B. In this study, a teacher (Ms.Cloud) tried to teach students who were having a lesson on probability using syllabus B. A math teacher educator, Susan, joined her and they decided to use math problem posing (MPP) to let students attach meanings of probability to real-life situations. The specific research questions were three.

S. S. Leung, *Mathematics Problem Posing in Action*,
https://doi.org/10.1007/978-981-96-8480-9_6

1. Is it feasible to develop MPP instructions for various topics in probability for vocational students in grade 11?
2. What outcomes were exhibited from problem-posing and problem solving (cognitive); and, acceptance (affective) from students?
3. What did the teacher suggest after development, practice, and reflection?

2 Background Literature for Actions

2.1 Probability Education

In history, studies on probability were categorized into three major phases: The Piagetian Period 2 (1950–1969), the Post-Piagetian Period (1970–1989), and the Contemporary period (1990–). The Piagetian period is dominated by research from Piaget, Inhelder, and other psychologists. They looked at children's probabilistic thinking and adults' strategies when predicting in uncertain environments. The Post-Piagetian period is characterized by Fischbein's seminal research on intuitions, decision-making under uncertainty, and probabilistic conceptions before and when students were instructed. The Contemporary period occurred after probability was in the school curriculum. Technology approaches could be used to show the distribution or law of the large number. This period was the main focus of this book.

Jones (2005) gathered work in elementary, middle, and high school and teachers from an international group of researchers. The purpose of the book is to build a cognitive and pedagogical basis to inform instruction. The introductory part was about him writing (by post at that time) to Fischbein and getting a reply within a month confirming the importance of his dissertation with young children on probability. He was touched by this response that confirm his study on young children's programs with theoretical and experimental activities. He remarked that people meet random events every day but tend to use deterministic lenses to evaluate the events.

In the handbook of research on mathematics teaching and learning, Shaughnessy wrote the differences between the perspectives of psychologists and mathematics educators. Psychologists mainly observed or described what happens when subjects solve tasks on judgments. However, mathematics educators developed tasks for teaching and learning and intervened to improve subjects' knowledge of probability. Perhaps, it is best if they work together to change students' misconceptions instead of having one-directional influences from psychologists' findings to implications for mathematics educators (Shaughnessy, 1992). More so, mathematics educators needed to work with school teachers to better their knowledge of teaching and learning and check if what was advocated in research can be applied in practice.

In talking about probability literacy, Gal (2005) mentioned that students learned probability for two reasons. The external one is that math and statistics is a field of knowledge. The internal one is that students learn to prepare for life, as chance and random events permeate our lives and environments. Gal (2005) included a concrete example in life situations such as upgrading insurance plans and thinking of the chance of risk for heart attack in future. Gal mentioned factors such as how people know about the world, personal dispositions, and the ability to analyze information given or implied. When students read a problem in probability (usually a word problem), they often encounter difficulty in imagining the situation, especially when the situation is not related to their experience. This understanding part needs great attention, not just on curricular inclusion or textbook problems using COVID-19 as a situation. It is a consideration that "*guides what gets planned, taught, assessed and valued in the classroom*" (Gal, 2005, p. 39).

2.2 Math Learning for Vocational Students

According to Ozdemira and Onder-Ozdemir (2017), knowledge of mathematics is important for each society because it "acts as a 'gatekeeper' to social progress" (Gates & Vistro-Yu, 2003, p. 32) and for a good profession. Open-ended questions were asked to 165 students about success and reasons why through structured interviews and classroom observation. Among 165 vocational high school students, 104 believed that they were unsuccessful or hesitant. The main reasons arise from students' perceptions of mathematics courses. Therefore, studies on making better math courses for students before leaving high school were needed. There were also complaints that many high school students do not have the math skills necessary for today's high-skill workplace or college entrance requirements. A study on teacher development was conducted to enhance mathematics instruction in five high school career-related programs (agriculture, auto technology, business and marketing, health, and information technology, see Stone et al., 2008). It was found that after a year of the math-enhanced lessons, students in the experimental classrooms performed equally on technical skills and significantly better than control students on two standardized tests of math ability.

Maybe the teaching of mathematics for vocational practice needs realistic situations. In vocational practice, the mathematics demands for most fields are very different from traditional mathematics taught in classrooms. Edo and Tasik (2019) developed high school teacher programs using a realistic math education (RME) approach. For example, they used species of sea turtles (members of sea turtle species found in nearby regions) in Indonesia and asked students to operate combined sets and examine the properties of set operation. It was found that the program proved to be successful in multiple measures of students' learning.

As above, there has been attention on probability education from Piaget till now, and indeed, all citizens need this literacy to cope with problems in life situations (Gal, 2005). In school, students were taught to solve "real-life" word problems. However, the situations may or may not be related to high school students' life experiences. What will happen when students are invited to pose a problem after solving one probability problem? Instead of reading the situation of a given problem, they provide situations about themselves. It is worth studying how vocational school students do problem posing in probability.

3 Method

3.1 Participants

The teacher, Ms. Cloud, was a devoted practitioner who was in graduate school studying part-time. She was trained as an elementary school teacher but also received qualifications for teaching vocational school. In teaching vocational school, she found her concern was to get low-motivated students not to dislike mathematics so much. "*A lot of learning problems that appear now are due to inadequate math competency in the elementary school.*" Finally, she decided to investigate using problem posing to teach a unit. Students were in Year 2 from vocational school (11 males, 23 females); their overall mathematics attainments were low. "*The students often query why solving math problems when were not meaningful.*" Ms. Cloud's class had syllabus B and did not include Bayes Theorem. Upon invitation by Susan, Mr. King followed the steps of Ms. Cloud and conducted the same typical MPP lessons that included the Bayes Theorem. The students were from an ordinary high school, also in grade 11, about 17 years old.

3.2 Design of MPP

She read carefully the materials and decided on probability. Ms. Cloud and Susan thought that probability was a good choice. In asking students to pose problems students will bring in situations with their real-life experiences. Figure 1, presents the structure of it in the curriculum (Syllabus A: 1-8; Syllabus B: 1-7). Item number 8 is Bayes Theorem.

As mentioned before, vocational schools have syllabus B while ordinary high schools have syllabus A. Bayes theorem was optional and Cloud did not teach it at all. This sequence in teaching this module is typical and is similar to all textbook series in Taiwan. To begin, teachers introduced classical probability, then students solved problems on characteristics of probability, conditional probability, and independent events. Example problems will be given in the part explaining instrumentation.

Ms. Cloud developed teaching materials so that she could teach by having students pose and solve their problems instead of solving problems from the textbook. Ms. Cloud's typical MPP class was as follows.

1. The teacher presented an initial problem
2. Students solved then posed a problem (same as an initial problem)
3. Students solved their posed problems
4. Students shared posed problems
5. Students completed diaries.
6. Teacher completed journals.

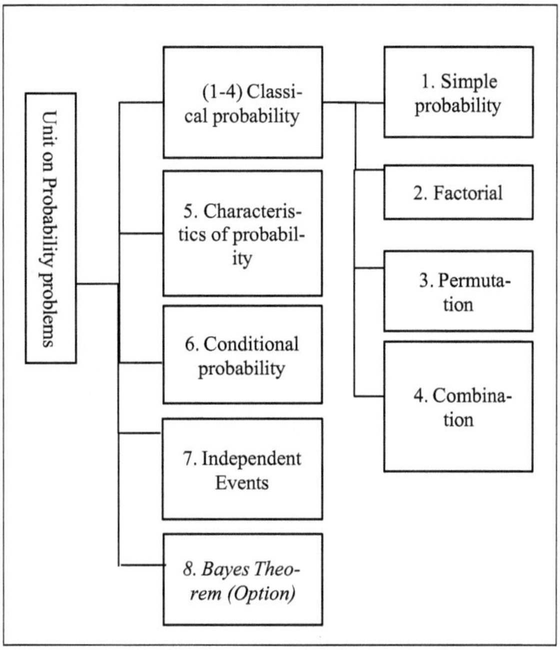

Fig. 1 Probability in math curriculum (Syllabus A: 1-8; Syllabus B: 1-7)

3.3 Data Collection and Analysis

Problem-posing worksheets, students' diaries, students' feedback teacher's journals, and audio tapes of interviews.

Problems were designed and went through a piloting. The MPP worksheets contained one typical problem on a unit. The materials were peer-reviewed by three teachers who taught for at least 10 years.

Unit	Name of Unit	Problem
1	Classical probability (simple probability)	The next-door neighbor, A-Zhu, is pregnant. What are the chances of A-Zhu pregnant with a girl?
2	Classical probability (factorial)	Today, Mei-Mei and Xiao-Ying participated in an interview with a company and found that in addition to them, 4 other people participated in the interview. The interview number is in order. What is the probability that Mei-Mei and Xiao-Ying's interview numbers are consecutive?

(continued)

(continued)

Unit	Name of Unit	Problem
3	Classical probability (permutation)	The toy store includes 5 kinds of dolls today (Stitch, Snoopy, Big Bear, Rabbit, and Sally) but there are only three display cabinets in the store: the window, the door, and the cashier. What are the chances of Stitch being placed in front of the window?
4	Classical probability (combination)	A group of 12 students including Xiao-Mei in Class A of the Catering Department of a higher vocational college. Three of them were selected for an intern at Delicious Restaurant. What were the chances of Xiao-Mei being selected?
5	Characteristics of Probability	According to the current epidemiological survey, 12 people out of every 1,000 adults suffer from colds, 15 people suffer from gastroenteritis, and 3 of them suffer from both colds and gastroenteritis. What are the chances of the teacher suffering from a cold or gastroenteritis?
6	Conditional probability	There were 30 students in Applied Foreign Languages (Class A) in a higher vocational school. 20 passed the English preliminary test, 10 passed the English intermediate test, and 7 people passed both. Mei- Mei passed the English preliminary test. What is the probability of Mei-Mei passing the English intermediate test?
7	Independent probability	During the Chinese New Year, department stores launched lucky bags. Department Store A had 50 lucky bags, one of which was a scooter. Department Store B had 100 lucky bags, one of which was a car. Mei-Mei went to A and B department stores to buy 1 lucky bag from each store. What is the probability of Mei-Mei getting both a scooter from department store A and a car from department store B at the same time?
8	*Bayes theorem (Mr. King's class)*	*There is a screening reagent for detecting an infectious disease. According to statistics, people with the disease have a 98% chance of being positive after screening; people without the disease also have a 2% chance of being falsely tested and being positive. Assuming 5% of the population in an area has the disease, and someone is known to test positive, find the probability that that person has the disease*

After each class, she analyzed students' posed problems, also the solutions, and opinions about the problem-posing activity. Reflected upon practice and completed journals. Data collection included problem-posing/solving worksheets, learning diaries, teacher's journals, and audio tapes.

4 Results and Discussions

The investigators referred to data and reported results on feasibility, students' posing and solving (cognitive), students' acceptance (affective), and teachers' reflection.

4.1 Feasibility of Using MPP Instruction in Teaching Probability for Vocational Schools

Ms. Cloud developed tasks for probability instructions (7 units) through problem posing. The class had no class on the Bayes Theorem. During the first meeting with Susan, Cloud told Susan her decision. "*It was an option. I find it challenging enough just to finish the must-teach topics. Let me forget Bayes.*" Three colleagues (also mathematics school teachers in the same vocational school) read her problems and assisted in making a final version and assured her that they were appropriate. Among the seven units she attempted, there was one that she found hard.

It was the unit on factorial (!). "*Susan, factorial is a term and what- it-is can drive my students crazy already. There is nothing creative one can pose using factorial. It goes well when I introduce it by definition, then I will show computations on typical examples.*" To Cloud, this unit was the most difficult to administer MPP. For the other 6 units, Cloud and her students did not express any difficulty and posed legitimate problems that were solvable (see the result given under 4.2). Hence, on the whole, the enactment of these MPP tasks for the teaching of probability through MPP was feasible.

4.2 Students' Learning Outcomes: Posing, Solving, and Acceptance

Problem posing-result 1: Types of posed problems
Coding was done according to Leung and Silver (1997). The types were five: Not-a-problem, Non-math, Implausible, Insufficient, and Sufficient. Altogether there were 238 problems posed and 230 of them were sufficient (96.6%). Seven of them could not be solved because of insufficient data (3 in simple probability; 3 in factorial, and

1 in independent probability). There was only one plausible problem, which was from the unit on conditional probability. Examples are given below.

1. Sufficient Data
   【independent probability】 : The Dragon Boat Festival is coming. A breakfast store sells 100 rice dumplings and 1 of them includes glutinous rice. B breakfast sells 20 rice dumplings and 1 of them includes mushrooms. Mr. Ming goes to A and B breakfast stores and buys one rice dumpling from each store today. What is the probability that Mr. Ming gets a rice dumpling with glutinous rice and a rice dumpling with mushrooms at the same time?
2. Insufficient Data
   【simple probability】 : Mr. Sing bought a new car, what is the probability that he bought a Mercedes? (Insufficient Data)
   【Probability involving factorial】 : A, B, and 28 people in the same class, what probability of continuous seat numbers? (Insufficient Data)
3. Implausible
   【conditional probability】 : There are 20 people in the Orangutan school. In the school, 10 people have gone to Africa, 10 people have gone to Australia, and 12 people have gone to both Africa and Australia. Mo has gone to Africa, then, what is the probability that he has gone to Australia?

All in all, judging from students' outcomes, this posing another problem activity was feasible for vocational school students. Although the teacher presented one particular problem which was prototyped for students to solve and although students in posing made up another similar one this MPP activity helped them to realize the structure of a problem (given conditions, goals). Ms. Cloud found that they improved in subsequent problem-solving (see also Chapter 2)

Problem posing-result 2: Difference(s) to initial problem
The problems posed by students were compared to those presented by the teacher (Ms. Cloud) and the differences were reported below. For example, there were three differences: number change, events change, and, contexts change.

~Number changed (Larger, smaller)
Students attended to numbers written in a problem and they made them sensible. For problems that were not creative in mathematics structure, at least there were changes to numbers. Students managed to change numbers in ways that made problems solvable. Therefore, in doing MPP students learned more about what made sense (meant solvable to those who attempted and reached a solution) and whether numbers were possible and sensible in real life. One example was, "*There are 20 glutinous rice balls in bowl B*" as compared to "*There are 50 glutinous rice balls in bowl A*" in the initial problems. Note that the time was near Winter Solstice when family members got together to enjoy bowls of glutinous rice balls together.

~Events changed to what they care about during that week
A problem with interviews was changed to an election. An initial problem with the names of 5 toys was changed to a problem with 4 brand names of mobile phones.

Ms. Cloud checked and realized that during that week, there was the election of officials for the photography club and friends were comparing phones. Also, attending interviews for positions was changed to auditions for soprano (alto) for a choir.

~Context
Cakes: cheese, cream, chocolate, coffee arrangements on shelves.

People: They used names of well-known figures in the country, and people in school (classmates, teacher) to make word problems. The real-life context was added to include people they admired (liked) or were familiar to friends.

Lucky draw: tickets and coupons in nearby department stores.

Exam timetable: arranging exam timetable for several subjects with specified conditions like times of the day.

Festivals: making rice balls for Winter Solstice, arrangement for dragon boat competitions, box and wrapping combination for Christmas gifts.

Interestingly, for the arrangement exam timetable, they wrote, "*Let us put mathematics in the third period then Ms. Cloud cannot be the invigilator.*" They learned that Cloud had an administration meeting during the tired period. It was just a joke among students. They were thoughtful in thinking of what they often came across. For example, their posed problems included going for a driver's license test, and screening interviews. They used these contexts to pose probability problems.

Among the seven units, the last two were more difficult and the first five were those students could handle. The more difficult the unit was the more similar (or close to identical) the posed problems when compared to the teacher's problems.

Problem-solving: Students solved posed problems correctly

As given in prior reports (Leung, 2013), when students solved their problems they solved without having great difficulty. They already understood the problem and followed later steps (Plan, Carry Out, Look Back) as given in Polya (1945). Overall, only 7% of the posed problems were not solved. Some examples were reported here.

The first is on independent events. Not knowing how many parts a cake was cut into nor the number of strawberries and that both events were independent, students treated both as simple probability problems with "1/n" as the answer. The second is on factorial computation, when students asked two persons to throw a die and asked for the probability of throwing the same number, students treated this problem as a simple probability problem and did not use factorial to compute. Third is a combination problem, the student understood the problem and solved it but there was a careless mistake in the final step and wrote the answer as "15/4" instead of $\frac{4}{15}$. Before MPP was implemented, the average class score in probability was 45% (45 out of 100). However, after the implementation of MPP, when students solved one typical problem in each unit (out of seven); posed another problem, and then solved their own posed problem they solved correctly 98% of their posed problems. The improvement from 45 to 98% was exciting to Ms. Cloud.

Students' acceptance of MPP
From students' diaries and feedback sheets.
"Impressive" part of MPP

Ms. Cloud's problem
The teachers' problem was the least among the list of impressive parts of MPP.

"Easy-difficult" level for the most part.

m-01: This problem is easier than the last one
m-25: 10000 times as easy!

"MPP processes"
Students expressed that in the processes of working together and doing discussions they had a lot of fun and feeling of achievement (m- 27: I can make a problem by myself, yeah!) and reminded themselves (m-13: I can fix the problem and make it solvable, felt so good).

"Sharing"
Sharing was the most frequent response among the list of impressive parts of MPP. They found sharing results from MPP interesting (m- 04: students could go to the stage to present problems) and had fun sharing (m-15: I have fun telling peers my problem. It was exciting and interesting).

English (1998) proposed that the levels of students' problem formulation should be divided into psychological level and sociological level. The psychological level includes whether students understand mathematical knowledge or relationships, use mathematical knowledge and relationships to figure out mathematical problems, and feel the solvability of the problem. In addition, the sociological level includes whether students strive to participate in group or individual exploration opportunities in the class, and identify themselves. Arguments engage in dialogue and share or critique created topics. In this study, it can be found that although the students have applied it to the psychological level, most of the impressions are on the sociological level. In particular, sharing and creating topics and participating in dialogues are the most feedback. It can be seen that peer interaction is a very important part of teenagers.

Responses to "How much do you like your problems?"
For 6 of the 7 units, there was less than 7% of responses showing they "disliked" MPP. There was one unit that 15% showed "dislike" (Unit 2, involving factorial). It was the most difficult unit for students.

Level 1: Like

1. Being able to make up a problem made me feel great. "m-04: *I fail in math for the most part but when I do MPP I can make up proper math, I admire myself! Wow, ha-ha*"
2. Creative and interesting (m-24) "*because I like to be creative*"
3. I can use what I like to make up the problems (m-23), "*I use the brand name of the phone I like to pose a problem*"
4. Using daily life examples to make up a problem (m-16-"*Sports Day is fun and I can pose many math problems on probability*")
5. "In MPP I think I know more math than before" (m-01)

Level 2: OK

Not sure: (m-28, m-19) "*I did not perform very well in MPP and my problems were ordinary maybe not up to standard; I can only think of problems that were similar to Ms. Cloud's*" (m-03; m-04). "*My problems are simple and not creative*" (m-04; m-05).

Level 3: Dislike

"*Not easy at all, except for the first unit*" (m-27). "*I do not think mine were great and my brain was absent*" (m-19).

There was one humorous student, "*I wrote something that would not happen, how could I get a "0" score? I just do not know how to score "0"*"" (m-17).

I. Diaries
II. MPP feedback sheets

① Like MPP: 28 Like, 6 Ok, 0 Dislike
② Did you prefer MPP? 31 preferred, 3 did not
Ms. Cloud's problems were close to students' liking, easy to understand, and helped them to think twice. They enjoyed the part of working together, more thoroughly doing discussion and sharing what they found. The lessons were lively, and relaxing and made them brainstorm. Those who liked what they used to express that Ms. Cloud was already great and gentle and they considered the usual way might be helpful for examinations.
③ MPP helped us to think: 17 Very much, 17 A bit, 0 Not at all. Overall they agreed that MPP helped but vary in degree of agreement. They expressed that MPP helped them to find the structure and logic of a problem.
④ MPP performance (Great versus need help)

Those who were great in MPP said they caught the problems' completeness, structure, and logic. Those who needed help in doing MPP presented that they had to learn to do reflections and edit extended problems.

4.3 Ms. Cloud's Reflections

4.3.1 About MPP Activity

(1) Beforehand: Find an appropriate teaching unit. Match the material to an MPP activity. Aimed at communication and kept records in the form of a diary, prepared to be ready to submit and share. In this case, Cloud did not teach Bayes Theorem and only 7 of the 8 units were attempted.
(2) Pilot and revision: After the pilot, the materials must be changed to what the whole class understands. For example, the original context of a textbook problem involved flipping one piece of sticks (to ask for the future) when going to a

temple. But not all students went to the temple and asked for the future. Ms. Cloud changed it to one that all students understood, "The next-door neighbor, A-Zhu, is pregnant. What are the chances of A-Zhu pregnant with a girl?"

(3) During MPP: Carried out MPP and taught probability as usual. Showed the structure of problems. Students studied carefully before posing problems. After doing MPP when one wrote the answer (of probability) which was greater than one the student would reject (e.g., "15/4"). The teacher encouraged them to ask when in doubt and advised on MPP classification standards; guiding them to check goals and conditions or requirements. The teacher must walk around; split the class into groups of two, and design worksheets to record results on problem-posing products and dairies to capture problem-posing processes. She read worksheets and diaries and selected the right person to go to the blackboard and present responses. The teacher tried to be humorous and lively acted promptly and attended to students when diaries were plain or had a section missing. The feedback that was immediately enabled students to improve rapidly.

(4) Afterwards: The teacher analyzed students' work and evaluated their teaching. That teacher responded to students immediately and positively changed students' willingness to participate in this activity. Arrange for students to boldly come to the front to present results. Watch the school calendar and plan activities accordingly. Pay attention to school events of the week: when there were important events of the week students lost concentration (e.g. Sports Day, English Week). The teacher might try to avoid clashes. This happened to Unit 2 and Unit 4. The participating students were exceedingly tired of both MPP and special events.

4.3.2 About Presentation of Students' Results

After 7 rounds of MPP instruction, presentation of results from activities, and collating ideas from worksheets, diaries, oral presentations, and small group work students learned by solving, posing then solving, fixing/editing problematic problems and realized that revising problems made the students learn a lot. In revisions, they learned why the posed problems were not solvable before they could revise or "fix" them. For example, "Mr. Sing bought a new car, what is the probability that he bought a Mercedes? (Insufficient Data). By adding the number of choices in the revision this problem could be solved. Even in nominating names of peers for a math problem, they learned to be fair and legitimate and avoided being dogmatic.

Wait time: It was not until the third unit then none responded that students disliked MPP. It took time to get used to MPP. The teacher must wait patiently.

Values in MPP: Students began to enjoy and like this MPP activity. And, they expressed that they learned the feasibility of complements of a problem. Not only could they spot mistakes in their own or peers' problems, but they could also find problematic problems in textbooks and exam items.

4.4 A Follow-Up Inquiry on Bayes Theorem

Susan and Ms. Cloud invited a teacher to follow an afterthought:

What types of problems will students pose on Bayes Theorem?

The initial problem:

There is a screening test for detecting an infectious disease. According to statistics, people with this disease have a 98% chance of showing positive after screening; people without the disease also have a 2% chance of being falsely tested and showing positive. Assuming 5% of the population in an area has the disease, and someone is known to show positive when tested, find the probability that that person has the disease.

A total of 23 students were completing the solving part. Only two did not solve the problem. One of the two students who missed the problem attempted with correct strategy but made careless mistakes in calculation. The response below was given by S28.

Mr. King's presented an initial problem. Students posed other problems (after solving the initial problem).

Out of 23, only 15 posed problems and wrote on worksheets. 10 students posed proper problems (category 5), 1 with insufficient data for solving (S11), and 1 which was not a problem (S6). The one with in- sufficient data for solving did not include the probability of telling a lie. The non-math problem merely relates that 90% of students failed to solve math problems. Interestingly, when students solved their posed problems, the two who could not solve were S11 and S6 who posed problems that were not possible to solve (insufficient, non- math). All the other students solved their problems successfully. This means that when the posed problem they understood students can find solutions successfully.

When asked if they liked textbook problems of the problem they posed and to provide a reason there were more preferring their own problems. The reasons were:

S28. The textbook problems were boring, my problem was interesting (about my idol).

S17. The textbook problems were plain, our problems were brainstorming.

S11. The textbook problems were simple, mine is wonderful.

S22, S29, S30. The textbook problems were too hard for us to understand, ours were easy to read.

One student said the textbook problem and his problem were the same, equally boring (S21). Finally, students' responses were shared and clipped to the back of the classroom. Mr. King compared the posed problem to the original problem he presented.

Context: Mr. King's students changed contexts to everyday real-life situations that were familiar to them such as buying concert tickets to visit Korean groups, the chance to pass a test to get a license, attending interviews for jobs, taking sick leave, injections for COVID-19, or, the chance of resulting burns and scald in the kitchen.

Numbers: Students included numbers that were close to the teacher's problem so that calculations would not be too hard. Representation. The problems were all written as text, without any tables nor figures. Finally, one student asked a problem on conditional probability instead of Bayes Theorem.

5 Conclusion

In the math curriculum, probability was often treated as a minor topic and not having enough importance as suggested (Jones, 2005). In Taiwan probability is often appearing as the last chapter in textbooks. As in the literature review, Gal reminded that: (1) its learning is a part of math knowledge, and, (2) it prepares students for life. Despite the aid of innovative tools and also improvements in technology, math teachers still faced challenges in teaching this topic and imagining students' intuition about chance. It is time for experts from math education and psychology to make a combined effort to work on teaching and learning (Shaughnessy, 1992). The investigators addressed this importance in a study on using MPP to teach it to vocational school students who may be entering the workplace soon. This study was conducted in Taiwan during the contemporary period after probability was explicitly given in the school curriculum (Jones, 2005). Students posing probability problems for high school and vocational schools.

It is feasible to develop MPP instructions for various topics in probability for vocational students in grade 11. Students were able to pose probability problems in designated types arranged by the teacher.

Favorable students' outcomes on problem-posing and problem-solving (cognitive); and, acceptance (affective) were exhibited. The results indicated that students were capable of doing so and were able to solve nearly all problems posed by themselves and their peers. Once the activity of problem-posing is the teaching approach students expected, they get used to MPP and gained experience like professionals. The most they enjoyed was the process of problem-posing and sharing posed problems with classmates. The investigators see that the values of MPP were not only on the psychological part in that participants improved cognitively over time but also gained values affectively and socially (English, 1998). In their opinion, the activity of problem-posing can stimulate the imagination. When teachers (Cloud, King, and Susan) implement problem posing, they discovered that students are fully engaged. Besides, they found the solvability and completeness of problems. In future, more attempts on other math topics in using MPP in high school can be investigated.

The team in this study made some recommendations. In teaching, teachers can implement MPP to probability in high school but also to observe needs and give immediate feedback. The attention is also given to include real-life problems not merely to attract interest but to aid comprehension of situations matching the topics in probability. This call is also for textbooks' inclusion of examples, as textbooks are meant for a large common denominator, not just for one school or one class. For example, the part on factorial is not easy for students to imagine. When there were such problems Ms. Cloud needed to explain factorial before doing MPP. For research, (1) Research can be extended to other high school students and students of other ages. (2) content: Other mathematical units may be considered, such as geometric figures, exponential functions, etc. (3) Try other kinds of MPP activities instead of "pose another problem" similar to the teacher's problem. (4) The structure of the proposed topic. (5) Discussion formats can be adjusted to the needs of students and the culture of the class. Finally, to respond to "*guides what gets planned, taught, assessed and valued in the classroom*" (Gal, 2005, p. 39) attention is for practice using research results.

Appendix: Data Collection Students' Diaries

Students' learning diaries

- About today's activity of Math Problem-Posing, what impressed you the most?
- Do you like the problems posed by yourself? Please check the following checkbox.
  ☐like ☐ordinary ☐dislike
  Why?
- Do you like activity of problem-posing today? Please check the following checkbox.
  ☐strongly like ☐like ☐ordinary ☐dislike ☐strongly dislike
  Why?

Students' Feedback Sheets

Feedback sheets of probability problem-posing

- Probability problem-posing (seven times) this semester. Do you like this activity?
  ☐like ☐ordinary ☐dislike
  Why?
- MPP, in contrary to our usual classroom teaching, which way do you prefer?
  ☐usual way ☐teaching with problem-posing
  Why?
- Did problem-posing assist you in solving problems?
  ☐helped a lot ☐helped a little bit ☐didn't help
  Why?

- Please recalled your attendance in class on problem-posing, chose one from four (A, B, C, D) and then explain.

  A enable me to think

  B enable me to improve creativity, different from doing examples

  C enable me to see of the problem is reasonable or not (whether it can be solved and described clearly)

  D enable me to reflect, stimulate creativity, then revised or ex- tended problems posed and solved.

  *The best thing that I did was (write code) ____, because____。

  *The part I should improve is (write code) ____, because____。

Teachers' Journal

Teachers' teaching journal

• education schedule • date of class • students' reaction and feedback • rethink after class

References

Batanero, C., & Álvarez-Arroyo, R. (2023). Teaching and learning of probability. *ZDM–Mathematics Education*, 1–13.

Chernoff, E. J., & Sriraman, B. (2014). Introduction to probabilistic thinking: Presenting plural perspectives. In E. J. Chernoff & B. Sriraman (Eds.), *Probabilistic thinking: Presenting plural perspectives* (pp. xv–xviii). Springer Science.

Edo, S. I., & Tasik, W. F. (2019). Design research on applied Realistic Mathematics Education (RME) approach in teaching math for vo- cational college. *Jurnal Pendidikan Vokasi, 9*(3), 294–306.

English, L. D. (1998). Children's problem-posing within formal and informal contexts. *Journal for Research in Mathematics Education, 29*(1), 83–106.

Fischbein, E. (1975). *The intuitive sources of probabilistic thinking in children*. Reidel.

Gal, I. (2005). Towards "probability literacy" for all citizens: Building blocks and instructional dilemmas. In G. A. Jones (Ed.), *Exploring probability in school: Challenges for teaching and learning* (pp. 39–64). Springer.

Gates, P., & Vistro-Yu, C. P. (2003). Is mathematics for all? In A. J. Bishop, M. A. Clements, C. Keitel, J. Kilpatrick, & F. K. S. Leung (Eds.), *Second international handbook of mathematics education* (pp. 31–73). Springer Netherlands.

Jones, G. A. (2005). *Exploring probability in school: Challenges for teaching and learning*. Springer.

Langrall, C. W. (2018). The status of probability in the elementary and lower secondary school mathematics curriculum: The rise and fall of probability in school mathematics in the United States. In C. Batanero & E. Chernoff (Eds.), *Teaching and learning stochastics. ICME-13 monographs* (pp. 39–50). Springer. https://doi.org/10.1007/978-3-319-72871-1_3

Leung, S. L. (2013). Teachers implementing mathematical problem posing in the classroom: Challenges and strategies. *Educational Studies in Mathematics, 83*(1), 103–116.

Leung S. S., & Silver, E. A. (1997). The Role of Task Format, Mathematics Knowledge, and Creative Thinking on the Arithmetic Problem Posing of Prospective Elementary School Teachers. *Mathematics Education Research Journal*, *9*(1), 5–24. Australia.

Ministry of Education (MOE). (2003). *The MOE deep cultivation project for nine-year curriculum and instruction.* http://edu.law.moe.gov.tw/LawContentDetails.aspx?id=FL029037&KeyWordHL=

Ministry of Education. (2018). *The general curriculum guidelines of 12-year basic education-Mathematics Domain.* Ministry of Education.

Ozdemira, H., & Onder-Ozdemir, N. (2017). Vocational high school students' perceptions of success in mathematics. *International Electronic Journal for Mathematics Education, 12*(3), 493–502.

Piaget, J., & Inhelder, B. (1975). *The origin of the idea of chance in children.* Routledge & K. Paul.

Polya, G. (1945). *How to solve it; a new aspect of mathematical method*. Princeton University Press.

Shaughnessy, J. M. (1992). Research in probability and statistics: Re- flections and directions. In D. A. Grouws (Ed.), *Handbook of research on mathematics teaching and learning* (pp. 465–494). NCTM.

Stone, J. R., III., Alfeld, C., & Pearson, D. (2008). Rigor and relevance: Enhancing high school students' math skills through career and technical education. *American Educational Research Journal, 45*(3), 767–795.

Commentary (Overview Chapter and Part I)

Lyn D. English

Chapters 1–6

The book, *Mathematics Problem Posing in Action: Students' learning, teachers' professional development and parental involvement* (Shuk-kwan S. Leung) is a welcome addition to the mathematics education library. Leung's admirable journey in fostering mathematics learning with and through problem posing is indicated in her first chapter:

> The book consists of actions taken in trying mathematics posing (MPP) in Taiwan. It is a sincere sharing from a mathematics teacher educator (MTE) who taught students and worked conscientiously with school teachers and parents over the 30 years of her career. The actions were taken as she assumed different roles in a research agenda on MPP.

Through her extensive range of research and professional development over several decades, Leung has had a major impact on mathematics education in Taiwan. She worked tirelessly with the mathematics education community to significantly advance the curriculum and students' learning and development.

Problem Posing for Teachers, Students, and Parents

As Leung points out, there is still much to learn about problem posing, from both teachers' and students' perspectives. Not only does Leung address problem posing from these perspectives, but she also highlights the important role that parents can play in children's mathematics development through problem posing. In her first chapter, Leung reviews several decades of research on students' and teachers' problem posing, highlighting the extensive work of Silver (e.g., 1994), Brown and Walter (1983), and of course, Leung herself (e.g., 2009, 2013). Importantly, Leung points out how problem-posing can occur naturally before, during, or after solving a

L. D. English
Queensland University of Technology, Brisbane City, QLD, Australia
e-mail: l.english@qut.edu.au

problem, with the actions of posing and solving being inter-related. Indeed, I anticipate there would be many instances where problem posing plays a major role, oftentimes without the solver realising it. As Leung has illustrated in her chapters, problem posing fits in seamlessly across the mathematics curriculum—and the benefits are far-reaching—provided quality tasks are developed.

Task Variable

Leung refers to "task" as one of the three major independent variables in problem posing, the remaining ones being "subjects" and "situations." In examining these variables, Leung illustrates the importance of the tasks presented to students to foster problem posing. In a previous commentary (English, 2019) I noted that the attributes of quality problem-posing tasks include a task's openness (open at task start, open at task end, or both), its mathematical demand and challenge, the opportunity for multiple solutions, and the development of creativity and flexibility. As Leikin and Elgrably (2019) emphasized, task openness, with the most open tasks (open at both start and end) enables flexible reasoning and creativity. Although Leung can see value in this form of task, she explains that she focused on tasks that could be readily used by elementary school teachers in her problem posing project.

Problem Posing for Enhancing Operational Learning

In the second and third chapters, Leung illustrates how learning of the basic whole number operations can be enriched through problem posing. Particularly inspiring is her detailed account of how a student requiring remedial assistance made substantial progress when given individual sessions involving problem posing with addition and subtraction. A pertinent point Leung makes is that textbook problems presented to students are already divided into sections on one single type with a focus on practice—followed by practice on another type (Leung, 1991). This is in contrast to the real world where a range of problems from different categories appear. As Leung indicates, there are several addition and subtraction problem types that can present difficulties for students, including change, compare, and combine examples. By engaging students in posing problems for these different operational situations, we can gain insights into their conceptual thinking and understanding, which we might otherwise miss.

Leung presents rich examples of how a teacher might adapt standard textbook examples on multiplication and division to foster problem posing. For example, *Multiplication: Set-up 1* and *Multiplication and Division: Set-up 3* are enticing and thought-provoking activities whose design can be applied to other examples. A case in point is an activity I implemented in second-grade classes, namely, *The Fun Park*.

Example of Problem Posing in a Group Activity

After a story about the invention of the Ferris Wheel, students worked in groups to make a schedule for a family of three to spend a day at the "Escape Fun Park." With $80 per group to spend, the children were to list the park attractions they wished to visit, together with food they would like to buy. Each group was given a Fun Park diagram with illustrations and names of a range of events/shows and various paths leading to the attractions. In making their schedule, the groups were to consider the starting times and expected duration for each activity, the costs, and the time it could take to reach each attraction they chose. The task required the children to choose how best to spend their money so that everyone in the family would enjoy themselves.

Video- and audio-recordings of group discussions revealed examples of problem posing and questioning as children worked on the task. After listing the attractions they wished to experience and their costs, the groups considered the times the attractions were being offered. The following excerpts illustrate some of the questions and problems one group proposed:

> *That (an attraction) is way too late. So, what do you have for your ideas, Annie? Maybe what should we do second?*
>
> *Could we go on the roller coaster?*
>
> *Space roller coaster...where is that?*
>
> ....
>
> *We all have the same thing. So it means nothing will be different in what we pay for. We are going to go all to the same places. Now let's see what should we be doing next?*
>
> *Let's just see how much money* (we have left). *Do you think we should do the grand parade next?*
>
> (Studying the given information) *That is actually at one-thirty. That is pretty late.*
>
> *We could go to that one last.*
>
> *What about the thunderbolt ride? That is at...9:30. That would work well, so let's do that one next.*
>
> *Space roller coaster is at one-thirty. And it's like the latest time that we should have gone. Lisa, what is your idea? Have you planned what you want to do?*
>
> *Grand parade next. What time is the grand parade?*
>
> *Three o'clock. But ... we have to eat first.*
>
> *Eleven thirty we stop. We need something until 12 o'clock.. So what should we do?*

After checking how much money they have left, one group member posed the problems of whether they each have $80 to spend or $80 in total and whether they all go on the same rides:

Yes, we plan this for all of us. (Fig. 2)

Moving to the remaining three chapters in Sect. 1, the richness of Susan's study is further illustrated in the treatment of the more difficult topics of addition /subtraction of fractions, ratio and proportion, and probability. Using her "Solve-Pose-Solve" approach, Susan indicates how grade 5 aboriginal students experienced success when problems were couched in contexts that were meaningful to them; they also

Figure 2 A group member solving one of the problems posed

wanted to choose contexts and problems that would be meaningful ("friendly") to the peers. Furthermore, Susan's findings indicated that through problem posing, students learned more about connecting math ideas.

I was especially interested in the reflective materials used in the lessons, including students' diaries, teachers' journals, and students "LIKE" sheets. Such materials can be readily adapted for other topic areas and for cultural differences. The materials could also include peer critiques, where students' peers provide feedback on others' problem creations. Another aspect worth noting in this chapter is Susan's emphasis on the importance of a unit, given that children's posed problems can lead to some confusion as to which unit is being referred to (e.g., type of fruit or weight). The examples Susan provided give a clear indication of the need for teachers to ensure this concept is well established.

Chapter 5 addresses the difficult topic of ratio and proportion. In reporting on her research with two teachers, Susan found that problem posing assisted the students in posing different problem structures, as well as number, objects, and contexts. In addition, there were differences in students' solving of related semantic structures and number choice in word problems. For the seventh-grade students, it was found that they also created "easy-difficult posing", where they progressed from posing problems that were easy to solve to those what were more challenging. The students' math diaries revealed rewarding feedback in that they were happy with their performance, enjoyed posing and solving problems, and increased in their understanding of the topic. Students also realized that they could use diagrams or drawings to represent their proportion problems.

The last chapter in this first section delves into problem posing in probability lessons in vocational schools. As a very real-world topic, probability enables the creation of appealing problems, albeit with challenges. Susan provides examples of a range of probability situations, ranging from classical probability to Bayes Theorem. Her study found that it is feasible to develop problem posing for various topics in probability for vocational students. With teacher facilitation, students were able to pose probability problems of designated types. Students displayed favourable outcomes on both problem posing and problem solving, as well as in the affective domain. Students were able to solve nearly all problems posed by themselves and their peers. It is perhaps not surprising that the students enjoyed most of all the process of problem-posing and sharing posed problems with their classmates. I have found this in several of my problem-posing projects (English, 1998). Students felt that problem posing stimulated their imagination, which was evident in their full engagement in the activities.

Concluding Points

As Leung has clearly illustrated, increasing both teachers' and parents' skills in capitalising on situations where problem posing can be fostered is greatly needed. Younger students display surprising abilities to reason hypothetically, pose problems without prompting, and consider multiple situations in posing and solving problems (Carpenter & Romberg, 2004; English, 2012; Lesh & Doerr, 2003). Furthermore, they enjoy posing problems! Problem posing can accommodate all levels of achievement in school mathematics, can readily relate to students' world, and can encourage creative and flexible thinking. As Leung illustrates in her book with numerous examples across topics and grade levels, learning and teaching mathematics through and with problem posing is both desirable and feasible—for all students at all elementary grade levels.

References

Brown, S. I., & Walter, M. I. (1983). *The art of problem posing*. Franklin Institute Press.

Carpenter, T. P., & Romberg, T. A. (2004). *Powerful practices in mathematics & science: Research-based practices for teaching and learning*. University of Wisconsin.

English, L. D. (2012). Data modelling with first-grade students. *Educational Studies in Mathematics Education, 81*(1), 15–30.

English, L. D. (2019). Teaching and learning through mathematical problem posing. Commentary for special issue, "Learning to teach mathematics through problem posing." *International Journal of Educational Research, 102*, Article 101451. https://doi.org/10.1016/j.ijer.2019.06.014

Leikin, R., & Elgrably, H. (2019). Problem posing through investigations for the development and evaluation of proof-related skills and creativity skills of prospective high school mathematics teachers. *International Journal of Educational Research*. https://doi.org/10.1016/j.ijer.2019.04.002

Leung, S. S. (1991). Building connections to the Pythagorean Theorem: An example of teachers' treatment of textbook problems. In M. K. Heid & G. W. Blume (Eds.), *Mathematics Education: Making connections* (pp. 41–47). Pennsylvania Council for the Teachers of Mathematics. [Handbook of PCTM]

Leung, S. S. (2009). Research efforts on probing students' conceptions in mathematics and reality: Structuring problems, solving problems, and justifying solutions. In B. Greer, W. Van Dooren, & S. Mukhopadhyay (Eds.), *Lieven Verschaffel* (pp. 213–225). *Modelling verbal descriptions of situations*. Sense Publishers.

Leung, S. S. (2013). Teachers implementing mathematical problem posing in the classroom: Challenges and strategies. *Educational Studies in Mathematics, 83*(1), 103–116.

Lesh, R., & Doerr, H. M. (2003). Foundations of a models and modeling perspective on mathematics teaching, learning, and problem solving. In R. Lesh & H. M. Doerr (Eds.), *Beyond*

constructivism: Models and modeling perspectives on mathematics problem solving, learning, and teaching (pp. 3–33). Erlbaum.
Silver, E. A. (1994). On mathematical problem posing. *For the Learning of Mathematics, 14*(1), 19–28.

Part II
Teachers' Professional Growth

Chapter 7
Building Professional Development Through Teaching Problem Posing for Teachers to Observe: A Math Teacher Educator's Storying on Equations for Three Grade Levels

1 Introduction

Chapter 7 through Chapter 10 in Section II presented cases on MPP and professional development. As given in Section I, an MTE (Susan) investigated students' learning using MPP. Here, she used problems in students' learning to do teacher training. For example, when students mix up the signifier from the signified in learning equations, she and her teachers will investigate "how to teach so that students learn to distinguish the two?" On MPP, the list of work for Professor Leung included: Trying to teach MPP as a school teacher for teachers to observe (Chapter 7); familiarizing herself with an updated curriculum, developing and shared MPP tasks (Chapter 8); and, reporting on a teacher as a curriculum developer (Chapter 9), and, on a teacher as an action researcher (Chapter 10).

When the teachers asked Professor Leung "How can we teach MPP?" She used cases on students' learning MPP to tell how. However, cases of successfully taught lessons and worksheets of students' problem-posing were not sufficient to convince teachers. The teachers' responses might be, "I cannot imagine how?" or "My students are different (unlike students you taught before), I do not think they can do MPP". Teachers asked how to set up tasks, and how to deal with responses from students. Professor Leung must take action—try teaching MPP. However, teaching MPP was brand new to this MTE. Professor Leung decided to borrow a class from one teacher and teach MPP in front of this teacher. She invited more teachers to observe, especially those who could not imagine how and said that their students were not able to do MPP.

When can this MTE teaching MPP made possible? How can she teach students and let their teachers observe? It worked well by fitting into the regular professional development sessions in Taiwan. In Taiwan, both elementary and secondary school teachers have common free time slots (no class to teach) so that they can attend teacher training sessions during regular work times. The time slots are fixed for certain school subjects (e.g., math, Friday afternoon). The sessions for professional

S. S. Leung, *Mathematics Problem Posing in Action*,
https://doi.org/10.1007/978-981-96-8480-9_7

development workshops could be for a whole school, for a district, and/or the state or even at the national level.

In this chapter, three lessons and observations on MPP were reported. The involved teachers were three: Math teacher educator, prospective teachers, and, in-service teachers. The follow-up discussion with the teachers in another classroom right after the class or in the methods course later constituted the salient part of her suggestion for teacher change using MPP as a means. The MTE asked three questions: (1) How did Ms. Leung set up and enact MPP tasks related to math contents; (2) What happened during in-class MPP teaching that required Ms. Leung to respond immediately; and, (3) How did the MTE use the sessions as a means to do teacher education. Three cases will be reported as stories, to suggest potentials in answering the above questions.

2 Background Literature for Actions

2.1 Professional Development for Math Teachers

Teacher education includes prospective teachers (PT) and in-service teachers (IST). In teaching, putting an idea down to practice is always a problem. For example, student teachers who can actively process knowledge may often have difficulties generating instructions consistent with current visions in the classrooms (Simon & Schifter, 1991, p. 272). In-service teachers, too, cannot imagine how; they often find training programs too theoretical (Lim, 2005; Sowder, 2007). Math teacher educators (MTE) are often considered to be those who voice from the ivory tower yet do not have experience in the ideas they suggest. Therefore, teacher education includes MTE and not just PT and IST.

Seeing is believing.

A way to solve this problem in practice is to show how; such as teaching students and teachers in the way you suggest them to teach (Krainer, 2004). For example, in-service teachers taking summer courses were stimulated to a constructivist view and changed their beliefs and decisions made in the classroom (Simon & Schifter, 1991). Therefore, having teachers try meaningful tasks first (e.g. to experience how mathematics can be learned in meaningful ways), may be a great approach to improve teachers' knowledge of the process of doing mathematics. Arranging for teachers to observe teaching or share demonstrations may also be helpful for teachers to imagine how, also for MTE to "see" problems arising in using MPP to teach a certain math topic (e.g. algebra).

There were other efforts in teacher education, such as to initiate change on the part of other teachers, to build cases, and to share exemplars in public. They included action research (Kemmis, 1991); participating in group projects on planning lessons on math topics (Ma, 1999), and publishing cases of teacher change in journals (Schifter, & Fosnot, 1993). These cases often take into account personal, social factors, and, cultural factors of math teachers around the world (Lim, 2005;

Llinares & Krainer, 2006). In a parallel vein, Taiwan also recognizes the need to develop a curriculum that considers its culture, local educational settings, and societal expectations in a curriculum change. In a reform of the curriculum, there was always a call for teachers' change. Taiwan's mathematics curriculum has been influenced by globalization trends. The mathematics curriculum reform initiated in 1993 referred to the NCTM mathematics curriculum standards (NCTM, 1989). And in each wave, the part on problem-solving remained as important as before. In the most recent two curriculum documents, and on the part of problem solving, math problem posing was explicitly mentioned (MOE, 2008, 2018).

In terms of professional development for math teachers, and, under recent curriculum reform in Taiwan, teachers needed to update on the difference in the sequence of teaching math contents (e.g., teaching early algebra in the elementary school) and approaches in teaching (e.g., problem posing to go with problem solving). Upon knowing that there is no need to wait until secondary school to teach algebra and that children can try posing (not just solving) math problems, what actions can teachers and teacher educators take?

2.2 *The Teaching of Algebra for Teachers' Observations*

Kieran (1992) provided an overview of the learning and teaching of algebra; how it used to appear as a highlight in secondary schools, and later on appearing earlier in the elementary schools. She also wrote an informative summary of algebra in the "Encyclopedia of Mathematics Education". In short, arithmetic is computational for the most part, yet, algebra requires the ability to use symbols and to describe relations. For arithmetic, just undoing will make you find the number inside the bracket but undoing does not always work in algebra for finding an unknown (Kieran, 2004). As noted, and as time went by, there was a move away from letter symbols and symbol-manipulation view to one that included multiple representations and also the vision of how algebra is learned. There is also a switch from just memorizing rules, and practicing equation solving to having students derive meanings for algebraic objects and processes.

The part on what research tells us about the learning of school algebra included successful approaches like the use of problem situations that are amenable to more than one equation representation. Also to engage pupils in comparing the two (or three) resulting equation representations to determine which one is better in that it is more generalizable. In line with multiple approaches, teachers can invite students to do problem posing and note the resulting students' learning outcomes. In posing problems involving setting up equations, do students have a sense of the letters and numbers represented in the word problems? In solving simultaneous equations, are students convinced that the solutions are correct? In setting up ratios in an equation, will the principle "rule-of-three" (Vergnaud, 1983, 1988, also see chapter 5 of this book) help students to realize when three of the four in the relation are given in a problem, one can solve for the fourth?

When students learn algebra, after they already learned arithmetic, do they think learning algebra is enjoyable? In addition to making this teaching-for-observation a professional development workshop, Professor Leung tried to show teachers how to teach math to make learning enjoyable. A questionnaire was given to 385 students (grade 1 to grade 12), asking if math learning is enjoyable (see Leung, 2012). Whether they think "yes" or "no", they had to provide reasons. There were four reasons: Whether math is useful or not, easy or difficult, interesting or boring, and, whether learning together or alone.

3 Method

3.1 *Participants*

3.11 The investigator and the teacher. The math teacher educator was the investigator. The teacher educator (Professor Leung) was also the person who stood on the stage (Ms. Leung) and taught MPP to students. There were three classes altogether. In each class, she told the organizer to introduce her as "Ms. Leung" (in Taiwan: *liáng lǎo shih*). *3.12 Teachers-the-observers.* They were teachers from an elementary school (in Hong Kong), a middle school (south Taiwan), and a high school (east Taiwan). *3.13 Students.* They were students (grades 5, 7, 10) of teachers-the-observers but taught by Ms. Leung by invitation. Ms. Leung faced students attentively and called them using gestures. The students answered right after Ms. Leung called and no one mistook her gestures.

3.2 *Design of MPP Materials*

Professor Leung was told which lesson to teach; she did not choose which math topic to teach. She used frequently asked questions to design her lessons. She set up MPP tasks and tried to make students' learning enjoyable: Whether math is useful or not; related to real-life situations or not; is easy or difficult, when math is interesting or boring, and, does math with friends or alone.

	Teacher's frequently asked questions
1. Equation with one unknown (Grade 5, elementary)	Students do not "see" what the unknown stands for in a word problem (signifier and signified)
2. Simultaneous equation (Grade 7, Middle)	Students do not know when is best to use elimination/substitution

(continued)

(continued)

	Teacher's frequently asked questions
3. The Law of Sines (Grade 10, High)	Students asked what was required to set up equations in a legitimate problem involving the Sine Law

The first case was about a student who erased the problem by mistake but the solution was there. The teacher invited students to make up a story problem matching "5x = 40…x = 8". The second case was on students-as-teachers posing simultaneous equations using contexts (bicycle and tri-cycle, chicken and rabbit, and combo for fried chicken). Teachers were "students" and had to solve them. The third case was posing one test item as a contribution to a coming test on the Law of Sine. In each case, Ms. Leung set up tasks to make math learning useful, easy, interesting, and, in a group setting.

In each case, there were two sessions: (1) Ms. Leung set up problem-posing tasks to teach while teachers observed; (2) Professor Leung (MTE) and teachers moved to a conference room and had a round table discussion. Students were absent during session two.

3.3 *Data Collection and Analysis*

From the teaching of math using MPP (first session): Video and photographs of lessons, and, worksheets of students. *From the professional development meeting (second session)*: Notes of teachers, focus group interviews, and, journal of MTE (Professor Leung). Data analysis was done by triangulating data and working towards: (1) How did Ms. Leung set up MPP tasks when she taught a given math topic? (2) What problems were arising that required Ms. Leung to respond immediately? (3) How did Professor Leung use artifacts of lessons to do professional development training for teachers and/or to prepare future teachers?

4 Results and Discussion

Three stories were used to present results on Ms. Leung's teaching MPP by setting up tasks for teaching a target topic that made students' learning enjoyable (See 4.1, 4.2, 4.3 below). In each story, there were instances where Ms. Leung responded to matters arising, and Professor Leung, urged Ms. Leung to take immediate action during the class and to do follow-up discussions during the professional development session. These three stories were used to answer the first two research questions (Set up and enact tasks, respond to in-class matters). The last research question (using the teaching/observation in subsequent teacher training) will be answered by reporting on Professor Leung at work (See 4.4).

4.1 Elementary School (5x = 40)

Session I. The target content was algebra for the fifth grade, aiming to teach students to understand the semantics of word problems. To set up and enact the MPP task, Ms. Leung drew a box on the blackboard. She said that the careless student erased the math problem but the solution was still on the chalkboard. Ms. Leung successfully aroused the curiosity of the students.

"Can you guess the problem?".

On the chalkboard, the solution was there for students to pose problems:

$$5x = 40;$$
$$x = 40 \div 5;$$
$$x = 8.$$

Professor Leung showed teaching skills (to teachers-the-observers) as she assumed the role of Ms. Leung. She patiently waited until the children fully understood what they were supposed to do. Students worked in pairs on the worksheet. Ms. Leung picked a student to orally present the posed problem. Preston read loudly.

> *Brian is 5 years old. Dad is "x" times as old as Brian. Dad is 40. Find "x".*

Leung also showed how to respond to in-class matters. Students were listening to Preston but not all students followed and understood. Therefore, Ms. Leung showed how she would check.

> *Boys and girls, watch. Ms. Leung will point to the chalkboard. Watch her finger and tell what she pointed stands for.*

$$5x = 40;$$
$$x = 40 \div 5;$$
$$x = 8.$$

Ms. Leung pointed to each hinted piece of information and asked what each piece represented. When she pointed to "5", students responded, "*Age of Brian, 5 years old*". When she pointed to "*40*", students responded, "*Age of dad, 40 years old*". Finally, she pointed to "x", and the response was, "*times, 8 times as old*".

> "Agree?"

Ms. Leung asked the one who submitted the worksheet. In this way, she showed the teachers how she handled responses from the floor. She turned to more worksheets and read out more posed problems. There were "5", "x", and, "40". She then loudly repeated the answers from the floor. Chocolate? Box? Person? Group? Individual? $? Ms. Leung took the trouble of constantly asking what the letter "x" and numbers represented in the posed problem. She called students to answer so that students could deepen their understanding of the meaning of algebraic equations. When one student finished, she called upon the next student. When students answered, she invited the problem poser to judge if the answers were correct or not. Students were sharing responsibilities in learning equations via problem posing.

According to Ms. Leung's design of the task (5x = 40), students could pose a variety of problems (*1-multiples, 2-isomorphism of measures, and, 3-Cartesian products*). When she read all the worksheets, only the first two types were given. "*I have $x, my brother's money is 5 times as much as mine, and my brother has $40, how much do I have?*" (*1-multiple*), and, "*One apple costs $5 and x apples cost $40 in total. How many apples are there in all?*" (*2-isomorphism of measures*)".

Professor Leung was curious to see if any guidance could make students pose a problem on Cartesian products such as the area of a rectangle. Her curiosity urged Ms. Leung to continue and to probe. She went to the blackboard and wrote one more line, "*Answer: 8 cm*". She hoped that the students would come up with an area problem.

She invited students to suggest more problems, "*If the student who answered wrote 8 cm, please guess this problem again?*".

Students discussed in whispers, and finally, someone raised his hands.

> *A piece of eraser is 5cm long, and, there are x pieces connected. The total length is 40 cm. How many pieces are there?*

However, the student asked, "*How many pieces*".

There was a need for Ms. Leung to respond to this matter. First, this problem was not in the area of a rectangle. Second, this posed problem did not meet the condition ("*Answer: 8 cm*") because the answer was "*8 pieces*" and not "*8 cm*".

Leung walked to the blackboard quickly and drew an eraser. She the added some dots on the right of the eraser indicating several pieces of eraser were on the right.

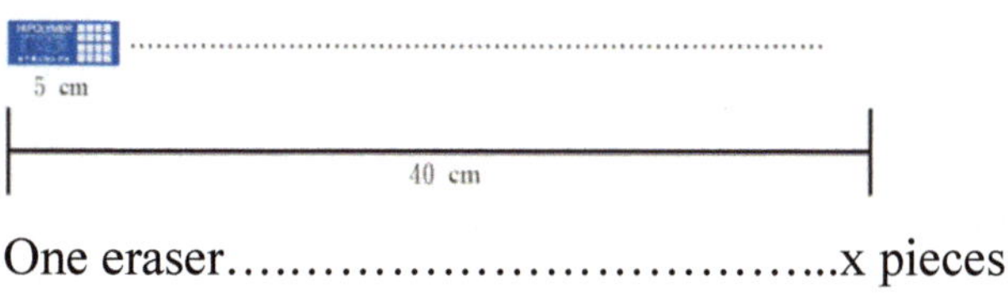

She spoke loudly, "*Hey, what is x?*".

Ms. Leung reminded students to check the units.

Professor Leung used this action to show teachers-the-observers there was no need to worry. When there was a mistake, the teacher could use mistakes to teach.

The "*Hey, what is x?*" drew the attention of both the class and teachers-the-observers. Some students remarked to the student who posed this problem. "*x pieces, but the answer was 8 cm?*".

The teacher educator was very happy. It was better for the students to spot this mistake than for Ms. Leung to correct students. Finally, Ms. Leung asked everyone to revise the work. The revision was:

> *A piece of eraser is x cm long. If you connect 5 pieces, the length is 40 cm. What is the length of one piece of eraser?*

The classmates said with satisfaction, "*8 cm long*". They revised accordingly and converted the posed problem to one having the answer as "*8 cm*".

The bell rang after class. Professor Leung's added "*Ans.* $x = 8$ *cm*" but students did not pose a problem belonging to the third category—The Cartesian product. Ms. Leung wrote in her journal to remind her what happened.

A problem like "*The side length of the rectangle is 5 cm, the width is x cm, the area is 40 cm*^{2}*. How many centimeters is the width?*" was not posed even though a prompt "*Answer: 8 cm*" was used as a prompt.

During the discussion meeting session (*Session 2*), she said if she would teach the next lesson, this area problem would be guided in the next class. Possibly, Ms. Leung would say, "*If your classmates come up with this problem, what do you think*?".

In this way, the addition "*Ans.* $x = 8$ *cm*" allowed the third type of multiplication and division word problem to be posed—the Cartesian product.

Session II. After class, Leung distributed the students' worksheets for teachers-observers to see students' outcomes. Grade five students expressed that the lesson was interesting, and found that math was useful, not difficult, and could be learned together (Leung, 2012). Professor Leung used the above in-service teaching training materials for training preservice teachers as well. They discussed about "setting up equation with one unknown" and "semantic types of multiplication word problems in the forthcoming "Math Teaching Methods" classes.

4.2 *Middle School (Simultaneous Equations)*

Session I. The target content was simultaneous equations for seventh grade. The set up was divided into three parts with different persons posing and/or solving.

Part 1. Ms. Leung posed. Students solved

There are 20 bicycles and tricycles, adding up to 50 wheels. How many bicycles? How many tricycles?

Students set up two equations with two unknowns and find the answers (10 bicycles, 10 tricycles).

Part 2.Students posed. Teachers and other students solved
Ms. Leung put up two cards (a rabbit and a chick). Students use heads and legs to pose. Ms. Leung walked around and gave me enough time to pose. When all students completed, she invited one to present the work.

I counted 38 heads and 116 feet. How many chickens? How many Rabbits?

Ms. Leung invited other students and all the 12 teachers to solve this problem. She read a solution from Mr. Mak (a math teacher at the back), as if Mr. Mak was one of the students.

18 chickens and 20 rabbits.

Ms. Leung then asked the students who posed this problem to tell if Mr. Mak was correct. The student sent a reply, "*Yes, Mr. Mak was correct.*"

Professor was about to go to Part 3 but she could not because Mr. Mak was not satisfied, "*You said I was correct. Why? Show me why I am correct?*".

The room was noisy.

Students were not knowing what to do next.

She pretended not to know what to do next. Ms. Leung smiled.

"*Can anyone help*?".

With this pretense that she needed help and some luck, the person who posed this problem nodded, "*Ms. Leung, please do not worry. I will show Mr. Mak*". She came to the blackboard and substituted the solution to the letters ("x" and "y"), and, to the left and right sides of both equations. This checking came into rescue. When both sides were equal, Mr. Mak was convinced that he was correct. In recall, Mr. Mak asked a very good question while being an observer and as well "a faithful student".

Part 3. Student posed and then solved
This time, the Kentucky Fried Chicken combo was the context. The picture included two options: (a) A coke with 6 pieces of chicken nugget, and, (b) A coke with 9 pieces of chicken nugget. Professor Leung's design was to address the question, *"How can we get our students to solve by elimination method. They just used substitution."* Students merely added the price for each kind of Combo and the problems were posed. Worksheets were collected and teachers at the back chose one for the whole class to solve. Guess what happened? Students used the elimination method to find the price of one piece of nugget. They did not use the substitution method. On the chalkboard, students read "6x + y" and "9x + y" on the left sides of two equations. Students compared. Unlike their habitual use of substitution, they preferred elimination this time.

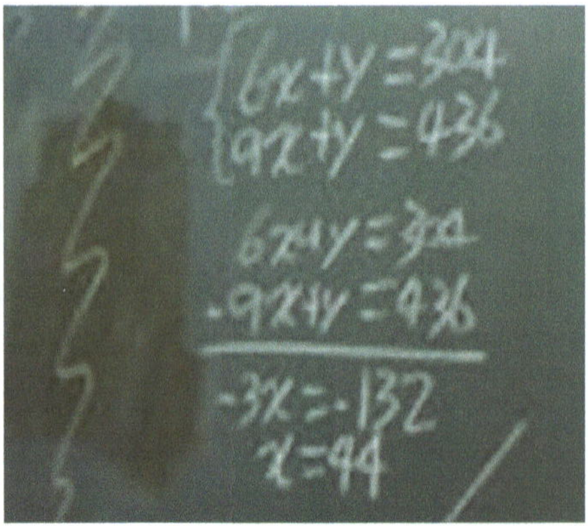

Session II: Teachers remarked that they learned to analyze students' work in groups for the first time. They remarked that they under-estimated students and were surprised at productive work on MPP. The teachers-observers noticed students paid more attention than before, as Ms. Leung taught. Most students expressed that it was the first time they made up problems. A teacher worried that students could not pose problems—that the teacher intended—but the rest said just let students try and even mistakes were not bad at all. Other teachers suggested allowing students to verbalize if they did not prefer to write.

Professor Leung took the actions of an MTE by using the actions of Ms. Leung during the last session. Together, they went through the lesson: Photos, worksheets, and small cards of students. Teachers were convinced that their students could do problem posing in simultaneous equations until they observed. Teachers-observers joined Professor Leung (Ms. Leung) to (1) analyze students' work and (2) reflect upon their practice.

Part 1. Teachers read and analyze students' work (n=30)

E.g. 1. Bicycles and tricycle.

Pose and solve: 15.

Pose but did not solve: 6.

Insufficient data to solve: 2.

Set up two equations: 7.

e.g. 2. Kentucky Fried Chicken combo.

The price of Combo A (one Coke and 6 nuggets) is $304.

The price of Combo B (one Coke and 9 nuggets) is $436.

One teacher read excitedly a posed problem she found interesting,

"*Tommy paid $1044 for 3 sets of Combo, how many are Combo A? How many are Combo B?*" Another teacher responded, "*Just substitute and you get the answers.*"

Part 2. Teachers reflect on their teaching after observation

Professor Leung, "*From a seminar in another school, the teachers told me that students prefer elimination and dislike using substitution, is that the same here?*"

Teachers agreed. "*Probably...but...will that depend on which two equations?*"

'*Today they used elimination when the coefficients of "y" in both equations are the same. They just minus and solve for "x".*'

Professor Leung purposely investigated how these practices supported teachers' reflections on observations and prior teaching.

From one of the teachers, "*It is good that we took the role of middle school students, and observed how opportunities can be created for our students when doing MPP, I can imagine what opportunities may arise for me to do the same one day...*".

Teachers expressed that there was magic in this problem-posing task.

Most students realized that "*to be a teacher was not easy at all*". Students also gave expressions of enjoyment (*Time flies! So happy! Very exciting, fun!*), thanks (*To dear Ms. Leung, you worked so hard*); and their learning (*After posing I understand more about simultaneous equations*). Teachers did reflections and recalled such instances, wrote to share with teachers in other schools or districts, and circulated through the network of the MTE.

4.3 *Senior High School (Equations Involving the Law of Sine)*

Preparation for Professor Leung. This arrangement affected the growth of this MTE. Professor Leung put in extra effort and went to practice chalkboard writing the day before. She used different colors and after some practice, the blackboard work met her expectations. The triangle ABC was inscribed inside a circle with 2R as a diameter and center at O. Angle A had the same measure as angle BDC. Sin A can be computed using the ratios of the opposite side and the hypotenuse of the right-angled triangle BCD (a/2R). Similarly, she drew two other angles (B and C) and used a total of three colors (red, yellow, and green) to mark the three right-angled triangles which were not easy to distinguish from each other when they were drawn in white inside the same circle.

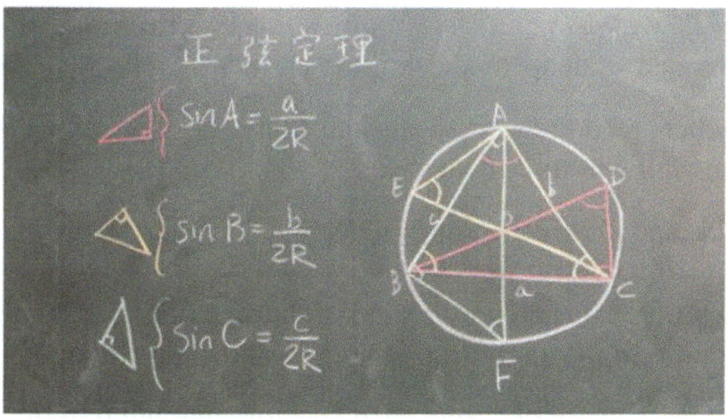

It was quite a long time ago since Leung taught the Law of Sine (grade 10). The teacher said that students were elite, compared to other classes of the same grade. Also, he taught the Law of Sine and there would be a short test on the next day. Leung intended to follow and give a review of this topic, "*Let me include problem posing. I will ask students to make up problems as contributions to the test.*"

Session I. The bell rang. The students greeted Ms. Leung first (Rise-and-Bow) and then turned 90 degrees to greet all the other teachers (teachers-the-observers sat by the side). Professor Leung arranged for observers to sit by the side so that teachers could see the faces of the students. When teachers observed they found that the students were very interested in having a new teacher. Accidentally Ms. Leung made a minor mistake in the chalkboard drawing. The chord BD did not pass the center (O), and she could not write "Sin A = a/2R". The students noticed and reminded Ms. Leung to fix it. The lesson went well despite this mistake.

Set Up-1 (Posing). "Now suppose that you are the teacher, you have to write exam questions that are related to the Law of Sine, and then give them to the teachers sitting next to you to solve them. You can pose any problem you want!" Professor Leung made sure teacher-the-observers watched the facial expressions of the students when they sat by the side. Leung used this MPP task to check whether the students understood these mathematical concepts through the posed problems. Everyone was very happy and was joking, "*It's our turn to pose problems for a test...*". Ms. Leung went to the floor, and the atmosphere was good. She walked around and picked problems one by one. Friends helped each other and then the whole class discussed the problems. The lessons went smoothly. Students were very excited that those questions might appear in the coming math test. They were "teachers".

Set Up-2 (Solving). Ms. Leung asked the students to work in pairs and test their friends. However, they only posed but did not solve them. Leung chose three items and pinned them to the blackboard.

These three exam questions said they were disappointed that solutions were missing.

The three were solved and Leung recorded the solutions. Then students solved all the other problems they posed. Ms. Leung discussed the problems that were not so easy to understand and asked how to re-formulate so that the items could be solved.

"*Imagine, if they appear in a test, and cannot be solved...what do you think?*" Ms. Leung reminded students to check if the problems were solvable.

Session II. Teachers read the worksheets and analyzed students' work.

The problems fell into three categories:

2.1 Given a side and a corresponding angle, find the radius (R) of the circumscribed circle.
2.2 Given the ratio of sides, find the ratio of the sine of corresponding angles
2.3 Given the ratio of the angles, find the ratio of the corresponding sides

"*Just now Ms. Leung chose one from each category and pin it to the blackboard for the students to solve*" One teacher-observed exclaimed.

An example for 2.1 was: *[side]* $A = 1368$, $A = 88°$, *find R.*

This student used the capital letter "A" instead of the small letter "a" to represent a side of the triangle. Ms. Leung resolved the confusion of using "A" to represent both the side and the angle. According to the convention, the problem should be written as, *"a = 1368, A = 88°, Find R"*. The choice of letters was made clear after Ms. Leung raised questions about them.

An example for 2.2 was: Given the ratio of three sides, find the ratio of the sum of two Sines to the Sine of the third sine of corresponding angles.

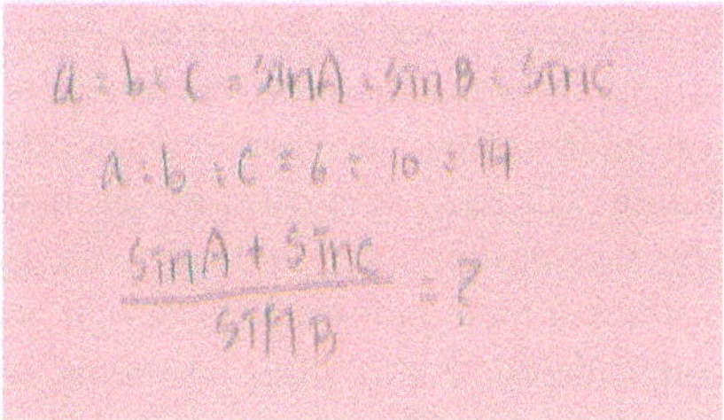

An example for 2.3 was: In triangle ABC, the ratios of the three angle measures are 3:4:5, find a: b: c (the ratios of the three sides).

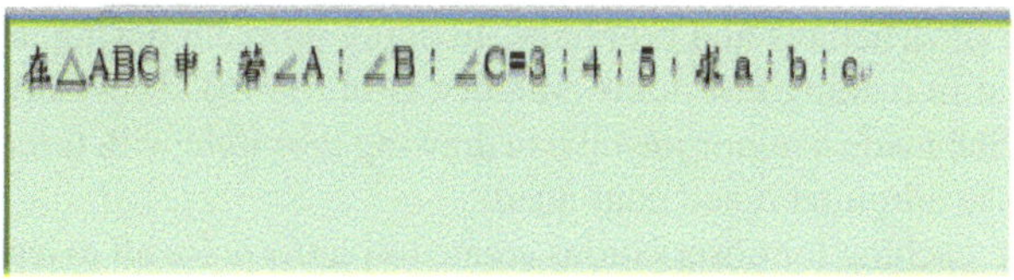

4.4 *Professor Leung at Work*

How were instances of lessons used in teacher training, professional development workshops, or future methods classes for prospective teachers?

In the first case (5x = 40), "See who can guess the problem that was erased from the blackboard?" successfully aroused children's curiosity. Prof. Leung attended to students' identification of letters and numbers used in an equation. Even when posed problems were legitimate and solvable, she purposely asked that each letter or number be represented in the word problems. In doing the checking back and forth, she liked to confirm the understanding of the signified and the signifier. This was what Prof. Leung wanted to address in the teaching and learning of setting up algebraic expressions matching word problems. She was patient enough to repeat checking of children understand what "5, 40, x, and 8" represented in the posed story problem. Both pre-service and in-service teachers analyzed the posed problems of the children and learned more about the semantic structures of word problems. The problems were mostly on multiplication and division. For pre-service teachers, the cases were used to show "how to do in-class math teaching" during the regular classes on Teaching Methods. They were excited to see Professor Leung teaching students, calling names, and selecting moments to respond using discourse techniques such as repeat, re-voice, explain, and, challenge (Bill et al., 1992).

In the second case, "students were teachers making up problems for teachers to solve" successfully aroused the interest of exciting students who attended classes taught by Ms. Leung (a visiting teacher). Students posed and teacher-the-observers solved. The chicken combo problem addressed a frequently asked question, "How can students learn when to use elimination and not stick to substitution." Prof. Leung also showed how to deal with unexpected responses and there was no need to answer questions that students might know the answers. The teacher who raised the problem, "Why am I correct?" was addressing a part that students often neglected: checking the solution by substituting "x" and "y" values to see if the balance still holds. Also, teachers realized the task design urged students to use the elimination method. Knowing when to use a quicker method is the key, though both methods worked.

In the third case, she considered "contributing test items" as a wonderful setup. In the lesson just before the text on this Chapter, this posed a problem for contribution to test items worked out well for the elite class. Students revised the lesson and they posed problems using "the rule of three" (Vergnaud, 1983). In using this principle, given any three, you can find the fourth. When students knew this rule, they could pose a variety of structures and problems. In doing problem posing, students could be professional in making test items (van den et al., 1995). They solved what they posed easily. She made a minor mistake in drawing the circle with three right-angled triangles but she admitted it and continued.

The way of teaching for observations created a platform for all to learn: In-service teachers, prospective teachers, and the MTE. They paid attention to professional noticing, including (a) attending to children's strategies, (b) interpreting children's understandings, and (c) deciding how to respond on the basis of children's understandings (Jacobs et al., 2010). In-service teachers saw how the students learn MPP. They exclaimed that they underestimated the ability of their students. Prof. Leung learned how to deal with in-class matters arising. Children made mistakes in MPP (8 pieces versus 8 cm, case 1, $5x = 40$). Professor also learned to deal with Mr. Mak's query ("Why am I correct", case 2, simultaneous equations). Finally, she corrected the chalkboard drawing when students spotted a mistake (case 3, Sine Law). Her teaching at the University did not require her to write on chalkboard. The drawing required for the Law of Sine made her worry. After this attempt, she changed how she taught pre-service teachers, and added the part on "chalkboard" writing in the teaching methods classes.

5 Conclusion

While MPP is known to be important, teachers expressed difficulties in making attempts to use them (Leung, 2013). Here, three stories by Professor Leung (an MTE) were given to show in-class MPP teaching. This MTE had not tried MPP before, yet she needed to introduce MPP to teachers. Whenever she could not convince teachers, she borrowed a class to try MPP. After she tried, she learned the pros and cons. Indeed, students, teachers (in-service, pre-service), and the MTE are co-learners (Jaworski, 2008). In this chapter, the MTE used MPP to teach (elementary, middle, and high school) on setting up equations (linear equations, simultaneous equations, equations involving the Law of Sine) and showed how MPP was used in building professional development. What did these teachers learn? On teaching MPP, set up tasks, in-class matters, and, building professional development.

On Teaching MPP. The teacher educator finally did math teaching. After a long period of not teaching grade 1 to grade 12, Professor Leung practiced task setup and in-class techniques for math teaching and in this case, using MPP as a means. Other than showing the observers possibilities, challenges, and difficulties (see Leung, 2013) she learned with teachers-the-observers, no matter who they were (in-service, pre-service). Teachers learned not to imitate the actions of Professor Leung but to bravely make attempts to try something new (e.g. MPP). They also read from these three stories and attach meanings to the three types of structure in problem posing (Stoyanova & Ellerton, 1996).

Structured: Given a solution ($5x = 40$, $x = 8$), find what the problem might be.

Semi-structured: Given a context (visual prompts), students then posed problems that involved the use of a particular method (simultaneous equations).

Free problem posing: Posing any problem on a specific topic, like the Law of Sine, and helping the teacher to test friends.

In teaching MPP teachers learn how students learn through MPP. They read the letter "x" and attached it to the meaning in a word problem when they posed and solved friends' problems; in turn, distinguished the part on signified and signifier. On setting two equations with two unknown equations, when asked "why this answer was correct" they learned to check answers by referring to both equations and not just checking the working of the calculations then believed the answers were true. On the Law of Sine, setting up the ratios and equating to 2R made students realize what was required; and how many givens were needed to find what was unknown.

Setup tasks. An interesting setup created a need to pose one problem. For the grade 5 class: Leung learned that these grade 5 students took turns to erase the blackboard. Ms. Leung said that the problem was erased and students were invited to guess it. For the grade 7 class: There were bicycles and tricycles on the school campus and students loved Fried Chicken, Ms. Leung added the traditional chicken and rabbit problems from textbooks. For the Grade 10 class: Ms. Leung pretended to be busy and asked students to make up test items on Sine Law. This was an elite class and they liked to challenge friends. Making test items is enjoyable for elite students. As given before, learning math through MPP was enjoyable when math was made useful, interesting, not too difficult or easy, and when students could learn together.

Respond to in-class matters. Ms. Leung knew that in-class teaching required constantly attending to and noticing students' responses (Jacobs et al., 2010). Unexpectedly, grade 5 students made mistakes in the number of units and the length of one unit in the eraser problem (8 pieces versus 8 cm). This unexpected mistake led the MTE to answer a frequently asked question about signifier and signified. Leung used it to show teachers the mistake was good for teaching children to match the unknown (or expression) back to the problem descriptions used in the word problem. Also, Grade 7 students said, "It was not easy to be teachers, posing is not easy." An expected response appeared: Mr. Mak asked why "18" was correct but Leung was calm enough and threw this question to the problem poser. The student (not Ms. Leung) used substitution to check both sides of the equation and convinced Mr. Mak. Leung's setup, students were teachers, was consistent for the whole lesson. She made the student who posed this problem answer Mr. Mak's question; students learned that it was not easy to be teachers. For the Grade 10 lesson on the Law of Sine, Ms. Leung's drawing was incorrect and she fixed it with the "help" of a student. The class was an elite class. Asking them to pose problems as suggestions for the next tests made the lesson enjoyable. Students listened attentively and reminded Ms. Leung that the diameter must pass through the center when the drawing was not accurate. In handling the in-class matters while borrowing a class to teach Ms. Leung (Professor Leung) co-learned with teachers-observers.

Build professional development. Let teachers observe MTE trying is a technique. In this case, the idea is on MPP. The worksheets were used in future methods courses for pre-service teachers. The artifacts of the lesson were also used for in-service teacher training. Video clips of discourse practices were created from this lesson. Grade 5, grade 7, and grade 10. Leung made a wrong diagram (the circle with diameter "2R" and sine a) but she corrected it and discussed it with the teachers. The MTE took this opportunity to remind teachers that there was no need to panic when a part of teaching was wrong. Admitting mistakes is part of the teaching. Professor Leung taught high school a long time ago. School teachers learned that they underestimated students. Prospective teachers learned through observations about how to respond to matters arising.

References

Bill, V. L., Leer, M. N., Reams, L. E., & Resnick, L. B. (1992). From cupcakes to equations: The structure of discourse in a primary math classroom. *Verbum, 15*(1), 63–85.

Jacobs, V. R., Lamb, L. L. C., & Philipp, R. A. (2010). Professional noticing of children's mathematical thinking. *Journal for Research in Mathematics Education, 41*(2), 169–202.

Jaworski, B. (2008). Mathematics teacher educator learning and development. In B. Jaworski & T. Wood (Eds.), *The mathematics teacher educator as a developing professional* (pp. 1–13). Sense Publishers.

Kemmis, J. (1991). *Action research for educational change*. Open University Press.

Kieran, C. (1992). The learning and teaching of school Algebra. In D. Grouws (Ed.), *Handbook of research on Mathematics teaching and learning* (pp. 390–419). Macmillan.

Kieran, C. (2004). Algebra. In S. Lerman (Ed.), *Encyclopedia of mathematics education* (pp. 27–32). Springer.

Krainer, K. (2004). On giving priority to learners' prior knowledge and our need to understand their thinking. Editorial. *Journal of Mathematics Teacher Education, 7*, 87–90.

Leung, S. S. (2012). Remedial teaching strategies for low motivation in learning mathematics. *Journal of Educational Research, 221*, 25–36. (In Chinese).

Leung, S. S. (2013). Teachers implementing mathematical problem posing in the classroom: Challenges and strategies. *Educational Studies in Mathematics, 83*(1), 103–116.

Lim, C. S. (2005). *Promoting mathematics teacher professional development through a school-based collaborative culture: Strength and weakness of the Malaysian and Shanghai experience.* Regular lecture, EARCOME3, Shanghai.

Llinares, S., & Krainer, K. (2006). Mathematics (student) teachers and teacher educators as learners. In A. Gutiérrez & P. Boero (Eds.), *Handbook of research on the psychology of mathematics education: Past, present, and future* (pp. 429–459). Sense Publishers.

Ma, L. (1999). *Knowing and teaching elementary mathematics.* Lawrence Erlbaum Associates.

Ministry of Education. (2018). *Curriculum standards for 12-year compulsory education-mathematics.* Ministry of Education. (In Chinese).

Ministry of Education of Taiwan. (2008). *Curriculum standards for national elementary schools in Taiwan* (In Chinese).

NCTM. (1989). *Principles and standards for school mathematics.* National Council for Teachers of Mathematics.

Schifter, D., & Fosnot, C. T. (1993). *Reconstructing mathematics education: Stories of teachers meeting the challenge of reform.* Teachers College Press.

Simon, M. A., & Schifter, D. (1991). Towards a constructivist perspective: An intervention study of mathematics teacher development. *Educational Studies in Mathematics, 22*(4), 309–331.

Sowder, J. T. (2007). The mathematics education and development of teachers. In F. K. Lester (Ed.), *Second handbook of research on mathematics teaching and learning* (pp. 157–223). Information Age Publishing.

Stoyanova, E., & Ellerton, N. F. (1996). A framework for research into students' problem posing in school mathematics. *Technology in Mathematics Education, 4*(7), 518–525.

Van den Heuvel-Panhuizen, M., Middleton, J. A., & Streefland, L. (1995). Student-generated problems: Easy and difficult problems on percentage. *For the Learning of Mathematics, 15*(3), 21–27.

Vergnaud, G. (1983). Multiplicative structures. In R. A. Lesh & M. Landau (Eds.), *Acquisition of mathematics concepts and processes* (pp. 127–174). Academic Press.

Vergnaud, G. (1988). Multiplicative structures. In J. Hiebert & M. Behr (Eds.), *Number concepts and operations in the middle grades* (pp. 141–161). NCTM.

Chapter 8
Provision of Math Problem-Posing Tasks for In-service Teachers and Analysis of Students' Outcomes for Prospective Teachers Using Examples of Data-Handling

1 Introduction

Knowing that students and teachers were inexperienced in MPP and did not perform well on MPP the next step was to work with teachers and to study how (Leung, 2009). When teachers were convinced and wanted to try, they needed example tasks to start. According to curriculum reform documents (MOE, 1993, 2008), students who learn math need to be actively constructing math knowledge instead of being passive recipients from teachers. A task requiring students to make up a problem, an MPP task, helps students to be active. In Chapter 7, teachers who observed teaching MPP began to realize that they underestimated their students and would like to try MPP. The question they most frequently asked hinted a direction for Professor Leung to follow.

Where do we get tasks suitable for teaching MPP?

Professor Leung decided to launch a project that included designing MPP tasks. In this project, she not only designed tasks; she also implemented the MPP tasks in the elementary classroom and invited teachers to create more tasks. There were three parts: *preparation, implementation, and, action research.* First, development of math problem posing (MPP) tasks aligned to the curriculum to help teachers try and start teaching math using problem posing. (*Preparation*). Second, involving teachers in using MPP tasks, analyzing children's posing mathematical problems, and using results in instruction (*Implementation*). Third, involve active practitioners to do and further develop MPP tasks to teach math using MPP (*Action Research*).

This chapter reports on the *preparation stage*, the beginning, and also the most important part of this project. This study aimed to provide MPP tasks for in-service elementary school teachers (IST) so that to have a head start, and for prospective elementary teachers (PET) to get ready for future teaching math using MPP in instructions. The research questions are:

S. S. Leung, *Mathematics Problem Posing in Action*,
https://doi.org/10.1007/978-981-96-8480-9_8

1. What *framework* was used to develop MPP tasks feasible for math instructions? (Math teacher educator, MTE)
2. What *student outcomes* did teachers collect and share? What *analysis and actions*? (In-service school teachers, IST)
3. When students' outcomes were used for training prospective teachers, what *pedagogical knowledge* was exhibited? (Prospective elementary teachers, PET)

2 Literature Review

2.1 A Need for Change in Teacher Education: Co-learners

Teacher training includes pre-service teachers, student teachers, and in-service teacher training (Sowder, 2007). In the past, professors (MTEs) were known to be more learned and would teach teachers how to teach, as in the "pipeline" metaphor). The teacher went to get "gasoline" from professors via a pipeline. Pipeline is one of five metaphors in research into practice given by Silver and Lunsford (2017). However, an MTE needed to learn again how to teach as they might have stopped teaching for a while. A framework, for co-learners sharing knowledge, was documented by Jaworski (2008). The model by Jaworski (2008) explained the mechanism when a Venn diagram represents how a mathematics teacher educator (MTE) and teachers share knowledge. One circle carries MTE's knowledge of research and theory while the other holds teachers' knowledge of students and schools. The sharing of knowledge, given in the intersection of the two circles, means that the passing of knowledge is bi-directional: teachers also pass knowledge to MTE. This sharing knowledge is a "cross border" metaphor given in Silver and Lunsford (2017); where researchers and practitioners learn from each other and work together for professional growth.

When a teacher educator introduces innovative ideas in MPP, she hears two frequently asked questions.

1. "What tasks can I use, the tasks in textbooks were on problem solving, not on problem posing?" or
2. "How can I find time? I have no extra instructional time for doing math problem posing".

Professor Leung referred to the above two questions. First, she provided MPP tasks for teachers to use. While there was a need for the provision of user-friendly MPP tasks for in-service teachers, there was also a need to introduce MPP tasks to prospective elementary teachers. Second, she decided to create tasks from grade 1 to grade 6 for all four strands (Number, Mensuration, Shapes and Space, and, data-handling) so that MPP tasks can be used by integration; requiring no extra time.

In all, this project can help involve participants (MTE, IST, PET) to know more about integrating MPP into the updated math contents in the elementary math curriculum.

2.2 *Development of MPP Tasks*

An alignment of MPP tasks to the math curriculum is needed so that there is no requirement for extra instructional time for teachers. MPP tasks for grades one to six were designed to follow the ordinary teaching sequence. The question Professor Leung asked was, "What types of MPP tasks would teachers use without excuse?" The answer is, "using a variety of math topics from grade one to grade 6 which means content-based." Also, "the MPP tasks should be structured yet open enough for problems to be posed."

1. Tasks with *math contents* aligned with the math curriculum, which also means following the textbooks.
2. Tasks with *a structure yet not completely open.* For elementary school teachers to use from grade 1 to grade 6. Six types of MPP tasks were used; and, named according to the information that was given (Leung, 2013).

2.2.1 Content-Based MPP Tasks

In this case, all designed MPP tasks were content-specific. The investigators were aiming at such tasks for instruction without extra instructional time.

2.2.2 Structure and Openness in MPP Tasks

As in Chapter One, MPP tasks are open, structured, or semi-structured. In this chapter, the target was on six types of MPP tasks, excluding free structured posing (Stoyanova & Ellerton, 1996) or given undefined and goal undefined (Reitman, 1965). For convenient purposes, and teachers' use in math instructions, a task was named according to the given (s), allowing students to refer to and pose a legitimate math problem that can be solved. The six types were:

1. Given an algorithm or expression (E)
2. Given a text in words (W)
3. Given a figure/table (FT)
4. Given a math topic (MT)
5. Given an answer (A)
6. Given a known problem (P)

In this literature review, the part of teachers' professional development deserves much attention on using partnership and not just having teacher educators to teach teachers. Projects with a common goal (e.g., integrating MPP into math teaching) and joint efforts were great for getting together to enhance students' math learning and for professional growth of teachers. Tasks designed, for trying MPP, are a possible common goal.

3 Method

3.1 Participants

Develop MPP Tasks. Participants were eight members of a university research team (Design of MPP tasks) for the development of MPP tasks. The MTE is the chief investigator of the research team of the project. One full-time research assistant was hired and stayed through the whole project (3 years); she was responsible for both research and administration. There were six part-time team members: graduate research assistants (GRA) who were elementary school teachers. The research team met every Friday afternoon.

Collection of students' outcomes. Participants were 105 teachers from ten elementary schools in the south of Taiwan. Ten schools, located in the same district as the university where the MTE taught, were randomly selected. Letters were sent to invite teachers' participation. The full-time research assistant for this project communicated with 105 teachers through a free district post service among schools in the city.

Pedagogical knowledge. Participants were 60 students who enrolled in teacher preparation programs at the university where this MTE taught, represented a group of prospective teachers (PET) who experienced content-based MPP tasks and students' outcomes for MPP. They took content courses and methods courses taught by Professor Leung.

Therefore, the MTE and seven team members, 105 school teachers (IST), and 60 PETs constituted the participants for this study.

3.2 Steps for Preparing MPP Tasks

The research team at the university designed MPP tasks with reference to grade levels, math contents by strands, and, types of MPP tasks. The tasks were in alignment with the math teaching sequence so that school teachers did not need to spare extra instructional time to use those MPP tasks. During the briefing, the MTE explained the rationale of the project and provided a framework for designing MPP tasks. The team met to discuss the designed tasks for problem-posing using the framework given by the MTE.

Professor Leung hoped that the tasks could help teachers to have a head start in their journey to MPP.

Names and reference numbers. First, there was a name and a number for convenient reference. The names were mostly the contexts (e.g., eggs, attendance, street lamp, or, read a chart). The number followed the order of the topic(s) appearing according to grades. However, teachers were free to use them anytime. For example, a grade four task might be used for grade three. Teachers could edit it to suit their purpose. A set of MPP tasks, formed a provision of tasks from grade 1 to grade 6, in all strands (from Numbers, Algebra, Space to Statistics).

Intended math contents. The team studied all textbook series used by the ten schools in this district ($n = 4$). All were reviewed before getting a license for publishing, making sure they met the requirements given in the official curriculum standards (MOE, 1993). The team members read problems from the four textbooks series and made a table of specifications after checking the common and differences of coverage and order of presentation of topics in chapters. A final table was made to indicate the contents and topics covered, following four strands, six grades, and designed in a context using one of six MPP types (see Leung, 2013).

Strands and grade levels. After the content was determined, there was a balance among strands and grades. There are four strands: Number and calculation (N), Mensuration (M), Shapes/Space (S), and Data handling (D). Within each strand, contents were listed from grade 1 to grade 6, using the names of textbook chapters. For example, under N (grade 1), there was "Count to 20"; under M (grade 3) there was "Reading calendar"; under S (grade 4) there was "Perpendicular and parallel" and, under D (grade 5) there was "Mean and mode". The table enables the team to check tasks in different strands and grades.

Structure and openness in MPP tasks. Please also read Chapter 1, for a consideration on Math Contents, Grade, MPP task type, and extent to which the task is open. Six types of MPP tasks were designed.

1. Given a text in words (W).
2. Given a figure/table (FT).
3. Given an algorithm or expression (E).
4. Given an answer (A).
5. Given a math topic (MT).
6. Given a known problem (P).

3.3 *Data Collection and Analysis*

Data collection and the formation of tasks were determined in a zip-zag path: Team members to and from school teachers. In the university, the six part-time research team members were divided into three pairs and were responsible for making up tasks according to grades (lower, middle, and higher). The set of MPP task drafts and revisions were made after Friday's meetings. The envelopes of potential tasks were sent to the ten participating schools for 105 teachers to read. When comments returned, the members discussed the problems that the teachers mentioned. After the meeting, three decisions were made: (1) accepted, (2) revised, (3) discarded (the problem was too serious and could not be solved even with revisions). Normally, the decision was "to revise". Then the envelopes were sent again to schools until tasks were accepted. The analysis was done in terms of suitability of MPP for math instructions.

Teacher's Feedback Sheet.

1. *Math Contents*: taught already ____, easy to be understood _____, include information that can be used to pose a math problem _____.
2. *Context*: familiar to school children _____, what context would you use otherwise_____.
3. *Language*: matching comprehension level for each grade_____, which grade would you suggest ______. Why?
4. *Notes on teacher's use* __________

Students' outcomes clipped below (Ref. Number of Task __).

4 Results

The teacher educator and team designed MPP tasks across six grades and four strands (Numbers, Measurements, Space and shapes, Data handling), and, across six types of problem-posing tasks. "D" is used for the strand on data handling, which includes statistics and probability. The products and process of the development are reported below. See Appendix 1 for an overview of 52 tasks.

4.1 Framework and MPP Tasks for Math Instructions (MTE)

4.1.1 The Final Set of MPP Tasks (Products)

With the combined efforts of MTE, research assistants, and school teachers, and after revisions and initial tryouts, the work was completed in a year. The final framework of MPP was developed based on (1) grade, (2) strands, and, (3) MPP types (Leung, 2013).

For each grade. There were six grades. The number of MPP tasks in each grade was about 9, and the exact number for grades 1 to grade six were 12, 6, 8, 8, 11, and, 7 respectively. *For each strand.* There are four strands: Number and calculation (N: $n = 23$), Mensuration (M: $n = 19$), Shapes/Space (S: $n = 5$), and Data handling (D: $n = 5$). *For each MPP type.* Given an expression (E: $n = 2$); Given a text in words (W: $n = 13$); Given a Figure/table (FT: $n = 25$); Given a math topic (MT: $n = 4$); Given an answer (A: $n = 6$); Given a problem (P: $n = 2$); and, Given a known problem (P: $n = 2$).

4.1.2 The Process of Developing MPP Tasks (Process)

As given in the Method section, the path is a zig-zag one with initial inclusion, then several stages in addition, revision, and deletion. The feedback sheets were thoroughly read and discussed during the regular team meetings. After several rounds of zig-zag paths of revision and reviewing, the decision was to accept items that successfully led to the posing of a variety of problems. Teachers referred to the classification of students' responses according to the sufficiency of problems (Leung & Silver, 1997). There were five categories: Not-a-problem, Non-math, Implausible, Insufficient, and Sufficient/Extraneous (Category 1 to Category 5). They tried the MPP tasks coded them, and then clipped one example of each category to the provided worksheet. Teachers' use of tasks by grade, strand, and MPP types and student outcomes were sent to the MTE by post. The following is an example of a task and students' responses (Fig. 1).

Task26.TV program

Time	Program	Time	Program	Time	Program
08:00	Taiwan Stocks	14:30	Daily Meal	18:00	Evening News
08:45	Hotline 168	15:30	Swallow Afternoon	20:00	2100 Public Talk
12:00	Noon News	16:30	900 Murderer	22:00	Shaking News
13:30	Sky Justice	17:30	Irresponsible Music	23:00	Late News

Codes/Description	Examples
1. Not a Problem	Mom and dad watched *Shaking News* at 22:00, I watched *Irresponsible Music* with my sister at 17:30.

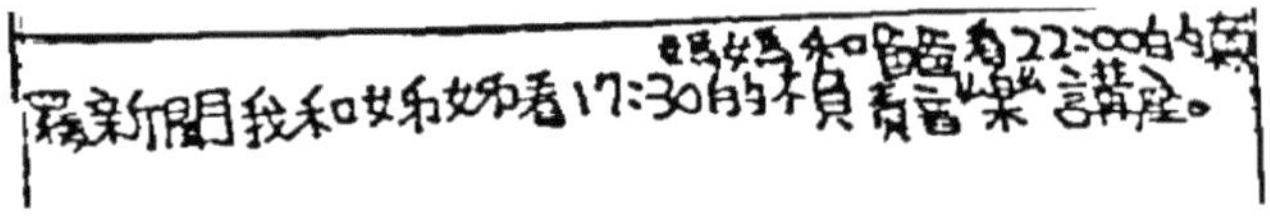

<This is a description, not a problem>

2. Non-Math	What programs are there on TVBS? Where can I read to know?

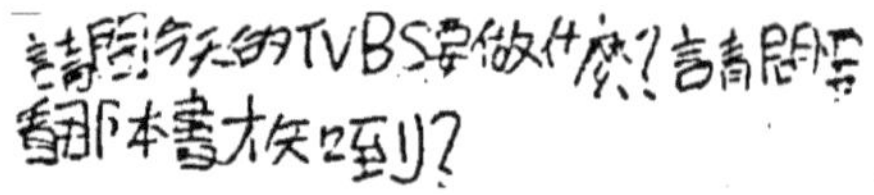

<This is not a mathematics problem.>

3. Implausible	In one minute I sold out 1/3 of a box of fruits today. From 21 o'clock to 7 pm, I sold what fraction of it

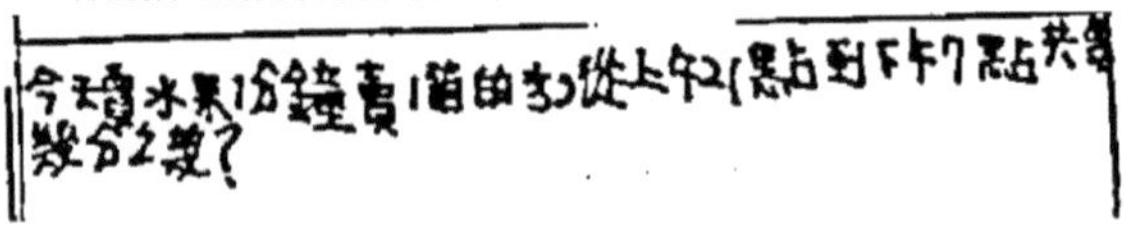

<The problem is implausible as the two sentences do not connect to make a problem.>

Fig. 1 Codes, descriptions, and examples of children's work

Codes/Description	Examples
4. Insufficient	Mom has to go to work at 12:00, which program can she watch? 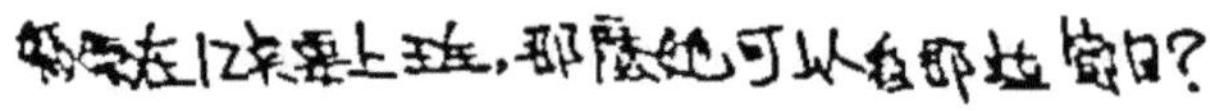 <*It is insufficient: The time to start watching TV is not given.*>
5. Sufficient	Today mom is not at home. You turned on TVBS and watched television from 15:30 to 20:00. How many hours did you watch? How many programs did you watch? 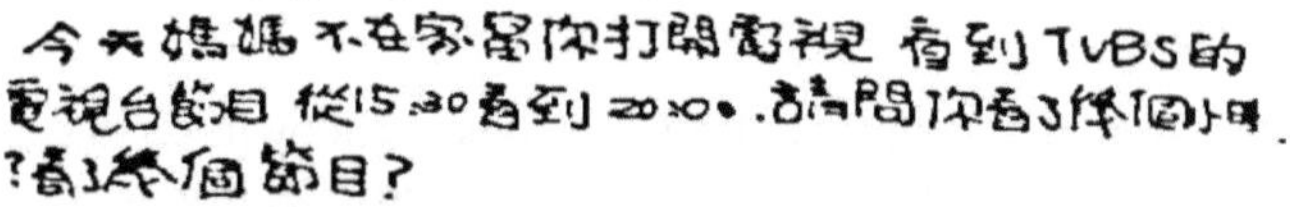

Fig. 1 Codes, descriptions, and examples of children's work

4.2 *Students' Outcomes from MPP Tasks Together with Analysis and Actions Taken for Math Instructions (IST)*

Although the teachers worked with MTE and team members when they tried tasks with students, the focus in this section is on the actions taken by in-service teachers (IST, 105 teachers).

Teachers examined the pack of MPP tasks and reviewed the appropriateness of *content*, *context*, and *language use*. Then they reported on the attempts to use MPP tasks, item by item. *Content-wise*, the final version of MPP tasks ($n = 52$) matched the math contents by grade and by strand; with teachers' flexible use of them across grades and strands. Examples of such flexibility will be given later on in 4.2 and also 4.3. *Context*. Contexts were revised using comments from practitioners. The contexts used were very much different from those used in the textbooks and a variety of representations (text, pictorial, math expressions, tables, figures) was used. *Language*. A greater number of revisions was made for grade one and grade two. The language used in the tasks was mostly pictorial. ISF suggested doing MPP orally for young children. Written work was tough for children at this age. For recording oral responses into written work, adults such as teachers and parents were involved. An emphasis on both the Chinese language and math usage would be helpful to students in the middle grade (grade three, grade four). For grade five and grade six, teachers suggested to train children to read math symbols and get used to the proper use of units for mensuration (e.g., km/hr. for speed).

Below were selected items, with teacher-returned sheets of students' outcomes. They also sent back their analyses and actions.

Below are four examples of task and evaluation.

1. Given a text in words (W):
   Perpendicular and parallel
2. Given a figure or table (FT):
   Eggs
3. Given a math expression (E):
   Division of fractions
4. Given an answer (A):
   Meter

There were five sample students' outcomes sent by in-service teachers (IST); one in each category (Sufficient, Insufficient, Implausible, Non-Math, Not-a-problem). The first example, *Perpendicular and parallel*, was used to show how the returned work from IST informed this MTE (professor Leung). From the information, the MTE learned (1) about the task, and, also learned (2) about the teacher(s). In each category, the student's work was given, then the teacher's analysis and actions followed. The MTE and team read and check clarity. They also checked to see if there was a need to talk with the teacher (individually or in group) meeting. During the meeting at school level, they made attempts to clarify the written work. The attention to the task and the teacher (s) was the same for the other three examples listed above; and, also to all the MPP tasks in the set ($n = 52$). Finally, selected examples were used in future teachers training session on MPP (district level or national level); and/or, for prospective elementary teachers who enrolled in teacher preparation programs.

1. <u>Given a text in words (W)</u>

Perpendicular and Parallel/grade 4/S/Item19

The MPP task was about "*Perpendicular and parallel*", which was meant for grade 4, strand S (Shapes and Space), with item no. 19.

The MPP task type is *Given a text in words (W).*

Below is the task description and five students' responses (*in another font, typed in italics*). The number in the parenthesis represents the identification number of the participating teacher, the teacher who sent back the students' responses (e.g. T2625). The part on analysis and actions taken, in response to each category of response, completed the presentation of the result of this particular MPP task.

Description of Task 19: *Perpendicular and parallel*
Please draw and then pose a problem about parallel or perpendicular.

Category 5 (Sufficient/Extraneous problem)
Read the figure, when lines A and B extend will they meet? When lines C and D enter A and B will they be perpendicular? How many were there? (T2625)

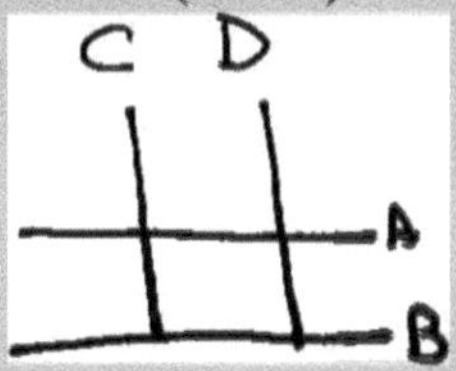

Analysis and actions taken

The lines given by students were not drawn by using a protractor and a straight edge. They appeared to be not parallel nor perpendicular. However, the teacher (T2625) thought it was okay and sent it back. The decision was to meet and discuss on this student's outcome. The team arranged to meet the teacher.

During the meeting, the teacher said that when no ruler and instruments was provided, she just accepted the students' handwritten answers. One teacher from another class disagreed. He said that he provided protractor for students to measure angles when they drew the figure to go with the problems they posed. "This task required a posed problem on parallel or perpendicular. If the figure is not correct. The problem will not meet the requirement." Another teacher also considered accepting this posed problem as problematic. She told T2625, "the two lines will meet as the lines are not parallel to each other. C and D will not be perpendicular to A nor B."

Category 4 Insufficient problem
perpendicular is 30 cm long, parallel is 69 cm long, how many cm altogether. (T2829)

Analysis and actions taken

The math concept and language used by the student were incorrect. This student thought of using length (cm) as a measurement of parallel/perpendicular; plus, adding the two did not make any sense. However, teacher T2829 sent this example back as insufficient (Category 4). After review in subsequent analysis from the team meeting, this problem was coded as implausible (Category 3) instead. If it is insufficient, it can be solved with the supply of missing information. However, this student's outcome cannot be solved, even after adding missing data. A follow-up meeting was arranged to meet T2829 to clarify the distinction between the two categories of students' outcomes. After explaining to T2829 the MTE provided more examples of implausible and insufficient examples. The team then invited T2829 to split the students' work into two types (implausible, insufficient) until he was able to distinguish the two. Another action was taken, to have teacher T2829 to interview this student. The student's vocabulary, on "vertical and horizontal", was mixed up with "parallel and horizontal". After the interview, the teacher knew that this student was asking about the total length of two-line segments. Through this action and checking guided by the MTE, the teacher (T2829) understood (1) the difference between implausible and insufficient examples; and, (2) that there was a need to interview the student to clarify written work.

Category 3 Implausible problem
My triangle board is a perpendicular one, how can I make a triangle? (T2421)

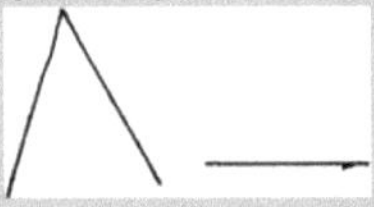

Analysis and action taken.

The teacher made a correct analysis. This problem cannot be solved, with both the given and the goal unclear. However, there was a need to interview students to clarify written work that was unclear to T2421. It was found that the student's "*My triangle board is a perpendicular one"* is a set square. The student wanted to ask, "How to draw a triangle using a pair of set squares?" instead of "*How can I make a triangle?*".

Category 2 Non-math problem
There was a pig, do you think its nose and mouth are perpendicular to each other? (T2819)

Analysis: no further action was taken.

This problem is a non-math problem and teacher T2819 made the correct coding

Category 1 Not a problem

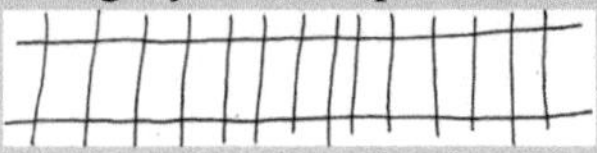

Analysis and no further action taken.

The student's outcome was just a drawing. This drawing is not a problem and teacher T2524 made the correct coding. (Category 1)

Actions were taken to either clarify, understand, or make suggestions to interview students. The process enabled the professional development of all participants over the school year.

Clipped below are the other three listed examples, presented with the inclusion of students' outcomes in each different category (coding).

2. Given a figure or table (FT)

Eggs/grade 1/N/Item 1

5	Mom bought 8 eggs. 2 eggs were broken. How many eggs remained?
4	The elder sister bought 8 eggs and some eggs were broken. How many eggs are there?
3	Mom bought 7 eggs. My younger sister took 2 eggs. She put them in the pocket and two eggs were broken. Later it turned into 6 eggs
2	Maruko bought 8 eggs. Some eggs were broken by Maruko's brother. Who broke the eggs?
1	Mother fowl laid 6 eggs and 2 were broken

(5) Sufficient/Extraneous; (4) Insufficient; (3) Implausible, (2) Nonmath; and (1) Not-a-problem

3. Given a math expression (E)

Division of fraction/grade 5/N/Item 30

$15\frac{3}{5} \div 3 = ()$

5	Now there are $15\frac{3}{5}$ L of salad oil, which should be divided into three barrels. How many liters are in each barrel?
4	There are $15\frac{3}{5}$ pieces of cake, and they are divided among 3 people. How much does the cake cost for each person?

(continued)

(continued)

3	$15\frac{3}{5} \div 3 = 5\frac{3}{5}$
2	$15\frac{3}{5} \div 3$ In the expression, there are two "3"s and two "5"s. Why is there only one "1"?
1	There are $15\frac{3}{5}$ acres of land for 3 people to grow rice

(5) Sufficient/Extraneous; (4) Insufficient; (3) Implausible, (2) Nonmath; and (1) Not-a-problem

4. Given an answer (A):
Meter/grade 4/M/Item 47
Answer: 25.6m

5	May had 29.9 m of cloth, and 4.3 m of cloth was used for weaving. How many meters of cloth May have left now?
4	There is a road that is 25.6 m long, how much is left unfinished?
3	A broken ruler is 21.1 cm long, and a clip is 4.5 gm by weight. How many centimeters and gm altogether are the broken ruler and the clip?
2	The teacher asked us to measure the distance between the classroom and the staircase along the corridor and then wrote on the blackboard that the distance was 25.6 m. Who did the measurement?
1	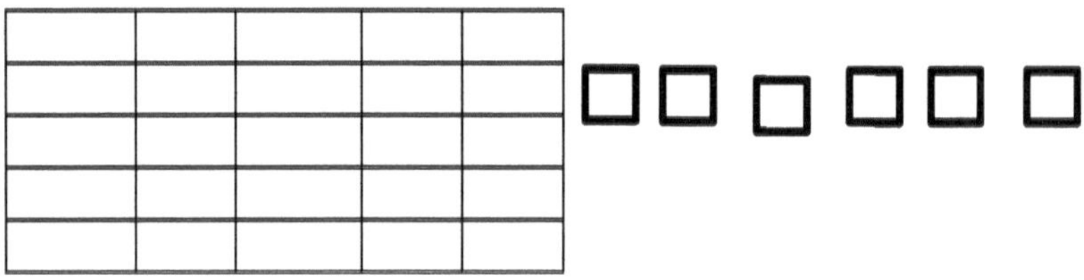

(5) Sufficient/Extraneous; (4) Insufficient; (3) Implausible, (2) Nonmath; and (1) Not-a-problem

Other than the above four, the set of 52 tasks with clipped examples of students' outcomes was printed into a resource book. It was used by the MTE to do teacher training. She used them for professional development sessions for in-service teachers (IST) and also for courses in the program for prospective elementary teachers (PET).

4.3 *Pedagogical Knowledge on Data Handling (PET)*

The focus of this section is on actions to study and unpack the pedagogical knowledge of prospective elementary teachers (PET, 60 students enrolled in a teacher preparation program). PET represented those who did not teach yet. They found it hard to imagine what students could and could not pose when given an MPP task. Therefore, analyzing students' work was helpful and practical for prospective elementary teachers (PET) in that they read what students wrote in a class using MPP. As an exercise for a course on teaching methods in mathematics, each PET analyzed at least one item (an MPP task from the set of 52 tasks used by ISTs). Each prospective elementary teacher read all students' responses and answered two questions:

1. Which grade would you use this item (Ref. No. _____)
2. What math knowledge of children can be inferred from the students' outcomes? What misconceptions (if any) were found?

Due to page limitations, an analysis of two of the MPP tasks (from 52) will be reported below. Both tasks with students' outcomes were from strand D (Data Handling). The first item is about *Money* given in a bar chart (grade 2, Ref. No.11). The second item is an *Attendance Record* of seven persons (grade 3, Ref. No. 48). The number of students' worksheets was 25 in each example.

4.3.1 Pedagogical Knowledge on Bar Chart (PET)

Money/Grade 2/D/Item 11

In Item 11, there is a chart with money ($1, $5, $10, $50, $100, $500, and $1000) given in the chart. The chart shows different numbers of each using rectangular grey boxes.

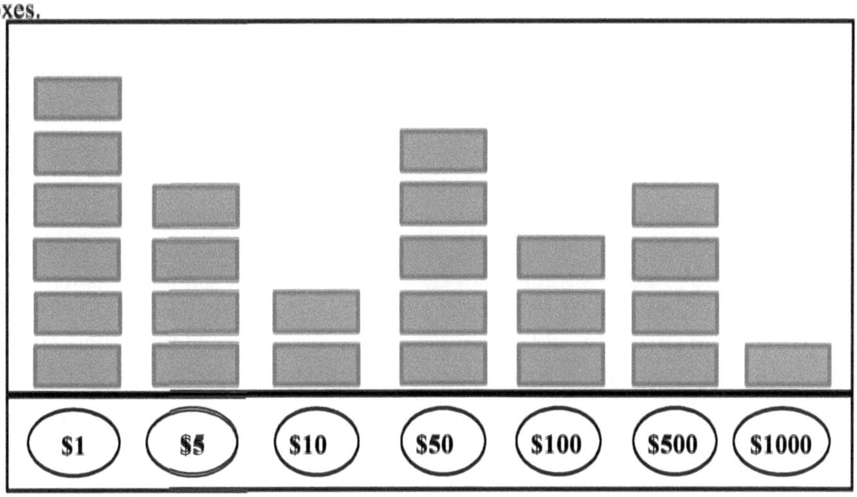

In grade two, children already learned about identifying bank notes and coins (Money, under the strand *Mensuration*). They also learn about whole numbers (under strand *Number*). In strand D (*Data Handling*), grade two children learn bar charts and how to read them.

Results of PET-1 analyzing this task.

<u>Grade Level.</u>

"*I will give this item to grade two children to try*"

The answer, grade two, was the same as in the textbooks and the math curriculum standards document.

Math knowledge or misconceptions

PET-1 found that children could use the chart to pose problems on (1) how many, (2) together, and (3) left over.
In the written work of PET-1, he expressed that the child learned about counting how many of each kind (money).

e.g. "*How many $50 are there altogether*?".

PET-1 found that children were able to ask about "altogether".

e.g. *"Dad had two $50 and mom had three $10. How many coins did they have altogether? How much altogether?"*.

PET-1 also realized that the child was able to ask about "leftovers".

e.g. *"If I took three $100 and spent $100 how much money was left?"*.

It was interesting to see that PET-1 appreciated the work of children at that age.

PET-1 exclaimed, "*As there was no need to solve, children might use counting to solve those problems. There was no need to do addition and subtraction.*" PET-1 also noted that although this task was meant for strand D (Data Handling) there was no instruction to require children to write about reading charts. Most problems were with reading the chart to tell "how many". There was none on comparing the information in the bar chart using "same", "more", or "less" or on finding which has the most (least). He also spotted that there were insufficient problems (*There are three $100, how much left?)* and one student's outcome that was in the Not-a-problem category (*Sorry, I don't think of any*).

Not being a teacher yet, he was happy to discover something about grade two children. PET-2 was excited to see that the amount of child's pocket money was disclosed through problem posing. PET-1 wrote, "*…it is interesting to see Pamela said her purse has…Now I can add and find how much pocket money she has kept in her purse so far…*" In this assignment, PET-2 was found to be enjoying reading about the real life of children. If not for the attempts of IST using the set of MPP tasks provided by MTE, they would not have experienced reading children's work so soon. PET-2 added, "*I look forward to my student teaching and entering the classroom.*"

In the methods class, Professor Leung asked for MPP teaching ideas from the PETs, such as how to guide more problems on this strand. She referred "*Sorry, I don't think of any*" and led a discussion.

"If children do not ask any, nor ask one about reading a chart, what can the teacher do?" said Professor Leung.

Some of the suggestions from PETs were:

"Let it be."
"I would ask a few and attract more"
"Which has the most number? $1"
"Which has the least?" "$1000"

"I will ask something to check if children are following such as how many $200 are there?"

The other PETs read the chart to see that there was no problem about "$200". They remarked, "That is a GREAT suggestion. I will try too...."

Then part of training prospective elementary teachers included knowledge of teaching and curriculum knowledge.

Professor Leung added, "*If this chart is for upper grades what can children pose?*".

One PET answered, "*Then children can ask ratio problems or percentage of one type to the total.*"

This example of one data-handling MPP task triggered lots of discussion in the Methods class for PETs. The MTE also used other tasks ($n = 52$) but in this chapter, only a few were used to report how she used them in preparing prospective elementary teachers.

4.3.2 Pedagogical Knowledge on Attendance Record

Attendance/grade 3/D/Table/Item 48

In Item 48, seven common children's names (Mei, Lan, Hua, Ling, Shu, Zen, and Jin) were given in a table as a record in a week. The table shows different persons on each of the days of the week except Sunday.

Mon	Mei	Lan	Hua		
Tue	Mei	Lan	Hua		
Wed	Ling	Lan	Hua		Shu
Thu	Ling		Hua	Zen	Shu
Fri	Ling			Zen	
Sat		Jin		Zen	

Results of PET-2 analyzing this task.

1. <u>Grade Level</u>
   This PET who analyzed this task read the textbook and curriculum, He said, "*This item, on how to do reading a table, was appropriate for grade 3.*"
   The reply, grade 3, matched the textbook series and the official math curriculum standards document.
2. <u>Math knowledge or misconceptions</u>

What did he think of math knowledge exhibited from all posed problems to item 48? To Professor Leung's surprise, he behaved as if was like another IST and coded them correctly. He read all children's outcomes and coded the posed problems into 5 categories. He finally selected them carefully and made the table below.

5	On Monday, there are three students on duty: Mei, Lan, and Hua. On Tuesday: Mei, Lan, and Hua. On Wednesday: Ling, Lan, Hua, and Shu. On Thursday: Ling, Hua, Zen, and Shu. On Friday: Ling and Zen. On Saturday: Jin and Zen. How many students are on duty this week?
4	In the table above, seven children were playing high jump in six rounds. Who lost in the first round? Who lost in the second round? Who lost in the third round?
3	Is the sum of all the names odd or even in this format?
2	The second group was divided into two groups. Group 1 had a group leader. Did the next group have a deputy group leader?
1	Mei's class was doing group discussion

(5) Sufficient/Extraneous; (4) Insufficient; (3) Implausible, (2) Nonmath; and (1) Not-a-problem

As noted above, PET-2, upon reading a variety of students' responses, could classify the posed problems into five categories according to Leung and Silver (1997). PET-2 did not write but listed the knowledge he thought children possessed:

Children learned from the table that different students took turns to share duties in a class.

The math contents related to the problems posed included reading and telling how many, also, adding up then considering odd and even.

The actions taken by PET-2 were made too. He proposed meeting the student posing a non-math problem. "*I do not know why the table invited a problem on deputy leader. If I were the teacher, I would ask the student why producing this outcome.*"

Professor Leung used this MPP task to do a discussion in the Methods class for PETs. She recalled a suggestion on the design of this MPP item. Compared to the previous *Bar Chart* item, this table presented information (names in Chinese) that were named. PETs suggested to use of geometric figures instead of names on this table. "*Names were not easy to read and compare… I already miscounted when I tried this task. I will try using circles, triangles, squares, rectangles, and so on…*".

The math teacher educator was pleased to see that the future teachers were exhibiting content and pedagogical knowledge when reading the children's work. They also appreciated the work and praised the children, "*There were not only posed problems with pure number computation but also some problems in context. Children were very smart.*"

The PETs commented also on the wording but had empathy. Both of them said as long as the math part was correct, they would "*check as acceptable*". As future teachers, they also attended to children's behaviors and conduct. When the context was not recommended, like, "*Student A was naughty and on average hit peers 86.4 times a week*", both PETs said, "If I were the teacher then I would find this child who posed this problem and talk to persuade him/her to change the story." "*Why not change it to helping the peers so many times on different weeks of the term then find the average?*" Leung was happy to note that apart from math education, PETs had a heart for educating the child to do the right thing to friends, to be helpful, and not to be a troublemaker in class.

Results on the provision of tasks and analysis of students' outcomes were used to make a resource book for subsequent teacher-training sessions. The book contained categories of children's work, and was also great in preparing PET.

5 Conclusion

In this chapter, Professor Leung (MTE) assumed the role of a bridge between in-service teachers (IST) and prospective elementary teachers (PET). There was professional growth for MTE, ISTs, and PETs in this project. All in all, the three types of participants were co-learners (Jaworski, 2008).

First, Professor Leung learned a lot as she supervised teachers. The metaphor of research into practice started as a *pipeline* metaphor. She developed a set of items ($n =$ 52) and showed teachers how to use MPP. The provision of tasks attracted teachers to start. The research team consisted of elite PET and IST heading for certification and/or graduate thesis on MPP. The teachers would make efforts to translate research to practice under the *translation* metaphor. During the process of walking over strands and grades on the curriculum map, reviewing the textbooks, and considering the nature of contents and grades, Professor Leung stepped in and out of the ivory tower as she led the team members. Later all of them began to exchange views and worked towards the final version of MPP tasks with consensus; on the kinds of information given in the MPP item (six types) and, on consideration of children's experiences to use contexts that were familiar and interesting to children of this age group. Over the long period of partnerships, the metaphor was from "*pipeline*" (having professors teach teachers), to "*translation*" and then later on to bi-directional metaphors like *osmosis*, and *border crossing* (see Silver & Lunsford, 2017 for five metaphors in addressing the gap between research into practices).

Second, for school teachers (IST), observing Ms. Leung's (MTE) teaching MPP was different from themselves using MPP actively in teaching. In this project, they learned what task was appropriate for whom (Silver, 2013). They also got a chance to read the results of her students who produced a variety of proper or problematic problems. Analysis of students' work was part of their professional development processes. They also learned to use results in MPP for further or remedial teaching. They were excited to see that their returned work was important, not only to themselves but to the future teachers in the teacher community.

Third, for prospective elementary teachers (PET), when they taught in the future, they wanted to get a head start in MPP. However, when they did not start to teach yet, how can they get experience and how can they imagine the "can" and "cannot" of children for each grade, and for a certain math topic? When they read returned work from IST and real children's work they had this opportunity. In this study, the chief investigator (MTE) used returned feedback sheets and children's work from IST, to compensate for the lack of teaching experience in this group of co-learners. There was also the affective side in preparing future teachers and on conduct. When PETs said, "*If I were the teacher...*" they would not only pay attention to math education, but they also needed to teach children not to disturb friends or to write properly in wordings to communicate with friends.

The initial part of a larger project launched by Leung was reported in this chapter. To help teachers started, a set of 52 MPP tasks, content-based, were developed. The materials were later used for in-service teachers' implementation (Leung, 2013) and for action research for three devoted teachers (Leung, 2016). In all, the project on developing tasks and analysis of children's work provided a chance for MTE, the group of ISTs, and PETs to learn together. Professor Leung was a bridge in teacher training who used both *one-directional and bi-directional metaphors* in getting research into practice (Silver & Lunsford, 2017).

Appendix 1: Distribution of six types of MPP Tasks by grade and by strand

Ref No	Name	Grade	Math content	Strand				Type of MPP tasks					
				Number (N)	Mensuration (M)	Shapes/ Space (S)	Data handling (D)	Expression (E)	Word (W)	Figure or table (FT)	Math Topic (MT)	Answer (A)	Problem (P)
1	Eggs	1	Divide and combine	N						FT			
2	Ducks	1	Compare numbers	N						FT			
3	Wood bars	1	Compare length		M					FT			
4	Stars	1	Count to 20	N									P
5	Knife	1	Measurement		M					FT			
6	Maru	1	Addition and subtraction within 100	N					W				
7	Liter	1	Addition and subtraction within 100	N				E					
8	10:25	1	What time is it		M							A	
9	Coffee	3	Measuring length	N						FT			
10	Balls, basket	2	Division	N						FT			
11	Money	2	What is it in the bag				D			FT			P
12	Crossing out money	3	Addition and subtraction (money)	N						FT			
13	Expression	3	+ - ×	N								A	

(continued)

(continued)

Ref No	Name	Grade	Math content	Strand				Type of MPP tasks					
				Number (N)	Mensuration (M)	Shapes/ Space (S)	Data handling (D)	Expression (E)	Word (W)	Figure or table (FT)	Math Topic (MT)	Answer (A)	Problem (P)
14	Bags of candies	2	Number and multiples	N						FT			
15	Calendar	3	Reading calendar		M					FT			
16	1854 is dividend	3	Number within 2000	N					W				
17	Pigs and chicken	2	A frog has four legs	N					W				
18	Units	3	Units	N					W				
19	Perpendicular and parallel	4	Perpendicular and parallel/ hour, minute, second			S			W				
20	Area	4	cm^2		M					FT			
21	Circle	4	Protractor and figure			S				FT			
22	Clock	1	What time is it		M					FT			
23	Capacity	4	Capacity and weight		M						MT		
24	Solids	5	Perpendicular and parallel			S				FT			
25	Mean, mode	5	Mean and mode				D					A	

(continued)

(continued)

Ref No	Name	Grade	Math content	Strand				Type of MPP tasks					
				Number (N)	Mensuration (M)	Shapes/Space (S)	Data handling (D)	Expression (E)	Word (W)	Figure or table (FT)	Math Topic (MT)	Answer (A)	Problem (P)
26	TV program	5	Timetable and time		M					FT			
27	Blackboard-1	5	Multiplication of fraction	N								A	
28	Factor and multiple	5	Factor and multiple	N							MT		
29	Revise a problem	5	Timetable and time		M						MT		
30	Division of fraction	5	Division of fraction	N				E					
31	Street lamp	6			M				W				
32	Pyramid and prizm	6	Pyramid and prizm			S							
33	Time and distance	6	Speed: km/h, km/min and km/sec		M					FT			
34	Calculator	4	Calculator	N						FT			
35	Fruit market	3	+ - ×	N					W				
36	Money	1	Paper money and coins		M					FT			
37	Mean, mode	5	Mean and mode	N					W				
38	Solids	2	What is in the bag				D			FT			

(continued)

(continued)

Ref No	Name	Grade	Math content	Strand				Type of MPP tasks					
				Number (N)	Mensuration (M)	Shapes/ Space (S)	Data handling (D)	Expression (E)	Word (W)	Figure or table (FT)	Math Topic (MT)	Answer (A)	Problem (P)
39	Trapezium	5	Formula for Trapezium area		M				W				
40	Flowers, bells	1	Compare	N						FT			
41	Sandwich	6	Surface area, volume of prims		M				W				
42	3-digit number	2	Number under 1000	N					W				
43	Speed	6	km/h, km/min and km/sec		M							A	
44	Plane figures	4	Km, m^2, m^3			S				FT			
45	Liter	6	1 L of water, 1000 cm^3 of wood blog		M						MT		
46	Read a chart	6	Percentage chart				D			FT			
47	Meter	4	Milliliter, km, m, kg		M							A	
48	Attendance record	3	How many days?				D			FT			
49	Bread	5	$4 + \frac{2}{3}$ and $4\frac{2}{3}$	N					W				
50	Temperature	4	Reading thermometer		M					FT			

(continued)

(continued)

Ref No	Name	Grade	Math content	Strand				Type of MPP tasks					
				Number (N)	Mensuration (M)	Shapes/ Space (S)	Data handling (D)	Expression (E)	Word (W)	Figure or table (FT)	Math Topic (MT)	Answer (A)	Problem (P)
51	Coach	5	Divide evenly, no leftover	N					W				
52	Money (1,10,50,100)	1	Units for money		M					FT			

References

Jaworski, B. (2008). Mathematics teacher educator learning and development. In B. Jaworski & T. Wood (Eds.), *The mathematics teacher educator as a developing professional* (pp. 1–13). Sense Publishers.

Leung, S. K., & Silver, E. A. (1997). The role of task format, mathematics knowledge, and creative thinking on the arithmetic problem posing of prospective elementary school teachers. *Mathematics Education Research Journal, 9*(1), 5–24.

Leung, S. S. (2009). Research efforts on probing students' conceptions in mathematics and reality: Structuring problems, solving problems, and justifying solutions. In L. Verschaffel, B. Greer, W. Van Dooren, & S. Mukhopadhyay (Eds.), *Words and worlds: Modeling verbal descriptions of situations* (pp. 213–225). Sense Publishers.

Leung, S. S. (2013). Teachers implementing mathematical problem posing in the classroom: Challenges and strategies. *Educational Studies in Mathematics, 83*(1), 103–116.

Leung, S. S. (2016) Mathematical problem posing: A case of elementary school teachers developing tasks and designing instructions in Taiwan. In P. Felmer, E. Pehkonen, & J. Kilpatrick (Eds.), *Posing and solving Mathematical problems: Advances and new perspectives* (pp. 327–344). Springer.

Ministry of Education of Taiwan. (1993). *Curriculum standards for national elementary schools in Taiwan* (pp. 91–132). Tai-Jye (In Chinese)

Ministry of Education of Taiwan. (2008). *General guidelines of grades 1–9 curriculum for elementary and Junior High School Education.* Ministry of Education (In Chinese).

Reitman, W. R. (1965). *Cognition and thought: An information processing approach.* Wiley.

Silver, E. A. (2013). Problem-posing research in mathematics education: Looking back, looking around, and looking ahead. *Educational Studies in Mathematics, 83*(1), 157–162.

Silver, E. A., & Lunsford, C. (2017). Linking research and practice in mathematics education: Perspectives and pathways. *Compendium for research in mathematics education* (pp. 28–47). National Council of Teachers of Mathematics.

Sowder, J. T. (2007). The mathematics education and development of Teachers. In F. K. Lester (Ed.), *Second handbook of research on mathematics teaching and learning* (pp. 157–223). Information Age Publishing.

Stoyanova, E., & Ellerton, N. F. (1996). A framework for research into students' problem posing in school mathematics. *Technology in Mathematics Education, 4*(7), 518–525.

Chapter 9
Teaching Through Development and Evaluation of Math Problem-Posing Program for Young Children: Ms. Wonder's Journey on Number Sense

1 Introduction

How can young children do mathematics problem posing (MPP)? A teacher, Ms. Wonder, attempted to answer this question through practice. At the time when this study was carried out, she was teaching under one of the curriculum changes in the country. Teachers were encouraged to be curriculum developers. Under the climate of this change, the call was on having students actively construct knowledge instead of imitating the teacher. She wondered how MPP can be used to make children active. When young children start to learn how to read and write; the role of language is important for the child to learn other subject domains (such as math). When a teacher invited young children to pose problems, are there challenges in language as well?

In Taiwan, an elementary class teacher taught the same class for two consecutive years and the teacher taught nearly all subject areas. When teaching students in grade one, Ms. Wonder already experienced the first year of implementation of the Nine-Year-Alignment curriculum standards. At the time of Wonder's investigation, the students were in grade two. She decided to teach math using math problem posing (MPP). Given that she wanted to see if including MPP made any difference, Ms. Wonder found another class as a comparative group. The two class teachers shared similarities in the number of years of teaching and the ways of teaching. The two female teachers often prepared lessons together. They talked about the children after class and daily. The only difference was that there was no problem posing activity in the other class.

> *"I think designing materials to teach MPP in my class (grade two) is possible. I know them very well. I have been their class teacher for one whole year. I had given thought and made a plan during the long summer break. Let me start and talk to a neighboring class teacher."*

Ms. Wonder was determined to make an investigation on MPP. She talked to Professor Leung, a math teacher educator, during one of the regular meetings. They came to agree on a study with three objectives: *teaching process, learning products, and, task and subject variable*

S. S. Leung, *Mathematics Problem Posing in Action*,
https://doi.org/10.1007/978-981-96-8480-9_9

1. Development and evaluation of the grade two MPP program.
2. Analysis of students' outcomes in MPP.
3. Difference (if any) in problem among three types of tasks and between two kinds of subjects. Difference (if any) in problem-solving scores between two classes (with/without doing MPP).

2 Wonder's Literature Review for Actions

2.1 *On Curriculum Standards*

The students in her class were the ones who experienced the implementation of the nine-year alignment curriculum standards when they were in grade one. This guide (MOE, 2008) emphasized the part on active learning; and also, to appreciate the work of others. Ms. Wonder read the guide and found that MPP could accomplish what was suggested; and was filled with joy. For example, the part on integrating multiple subject domains. It was made possible in problem posing. In grade one and grade two, language and math were the two most important subject areas to prepare lessons for most teachers. The guide also made suggestions to the teachers and encouraged teachers to become curriculum developers. Wonder believed that she was creative and decided to take action.

2.2 *On Mathematics Problem Posing*

Ms. Wonder was reading Tsubota (1987), a translated book on problem posing written in Japanese. The book included example cases in grade 1 to grade 6. "It's a fascinating idea to work on problem posing. Let me carry out a careful study on my children." The next teacher who wrote about teaching K-2 problem-posing was from Australia (Skinner, 1991). In this book, the teacher shared her experience over years of teaching young children. There were examples of tasks and children's responses. Hence, the books by Tsubota and Skinner, convinced readers that young children not only posed problems but also solved and freely contributed ideas to friends. As Ms. Wonder read, she found the two books helpful to teachers in dealing with matters arising. For example, the young children timed the walk of an ant, when they could not solve a problem like, "*How fast an ant can run?*" (Skinner, 1991). Ms. Wonder also learned that it was a great idea to let young children verbally reported the problems. When in need, the teacher taught (or friends helped) them to record the posed problems in written form. Thus, MPP links to language use.

2.3 On Number Sense

Content-wise, Ms. Wonder also read literature on addition and subtractions, which were the same references Ms. Glory read for remedial teaching with a second grade (e.g., Carpenter et al., 1996, Heller & Greeno, 1978, English, 1998, given in Chapter 2 of this book). Therefore, this literature part on addition and subtraction word problems will not be repeated here. However, Ms. Wonder prepared to learn more about MPP with the teaching of MPP in number sense. In Skinner's book, the first four chapters report the "how to do" part on MPP and the last four chapters were on numbers: understanding, investigating, learning number facts, writing and recording number sentences. It was amazing to see mensuration examples like "*How big is your smile*" and, also with real-life contexts in the southern continent (Australia). Wonder agreed that cultural contexts are interesting to young children. She decided to use Chinese cultures like Moon Festivals in her materials for MPP.

Next, she read a report (Leung, 2013) and came across various types of MPP tasks. There were six altogether. *"My class is now in grade two. There is no need to use all six types. Let me choose a few types that were appropriate to grade two."* After some thought, she chose three types: Drawn, text, and math expression. The other three (answer, problem, and solution methods) could be used in upper grades.

Finally, she read the thesis written by a prior investigator under the supervision of the same advisor. The thesis wrote about the choice of math topics after teaching MPP for a whole year round on all topics (see Ms. Kitty, Chapter 3), *"topics with rich contexts and connections to other math topics are best for problem posing."*

Ms. Wonder told herself, *"I am ready. Let me begin?"* Professor Leung worked as another investigator, accompanied her, and documented Ms. Wonder's journey as a curriculum developer.

3 Method

3.1 Participants

The teacher. Ms. Wonder was a warm and gentle colleague in an elementary school and as well a popular teacher to young children. Wonder represented an elite teacher. She graduated from a teachers' college that prepared elementary school teachers. After attending an intense screening procedure on academic performance in high school and interviews she was admitted to this special teacher program. The four-year college program was tuition-free. Upon graduation, she was allocated to a school to teach and did not need to look for a job. And, to which school depended on her overall performance during the four-year teachers' training program. All in all, she represented one of the elite students in the college who worked very hard. Finally, she was happy to know that her dream, to teach in a school near her home, came true.

Students. There were 36 students in Wonder's class. Students in this school enrolled according to home addresses. Their general performance was about the same in entering this school at grade one. That was to say, each class in each grade was about the same in academic performance. A normal distribution was the description when asked. The community nearby was a bit old and their parents' occupation was mainly workers. Students were having school under the climate of the first-year implementation of nine-year alignment curriculum standards (MOE, 2008).

3.2 *Design of Materials for the MPP Program*

In Taiwan, a class teacher was free to use her instructional time as long as she followed the country's curriculum standard documents and the textbook of the school. At the time of the study, Ms. Wonder took charge of her math teaching and designed a special MPP program for her class.

3.2.1 Stages and Duration

In each week, there were three math lessons. In this particular MPP project, she spent two of the three math lessons on MPP per week. There was a total of 12 weeks and 24 sessions. Ms. Wonder and Professor Leung met and made decisions. MPP was a completely new activity for both Ms. Wonder and her students. The implementation of MPP was done gradually. Ms. Wonder's program was divided into four stages: Stages: I, II, III, IV (with 2, 3, 3, and 4 weeks respectively) adding to a total of 12 weeks.

I. Oral Posing (2 weeks)
II. Written Posing (no need to solve) (3 weeks)
III. Pose then solve your own problem (3 weeks)
IV. Pose and solve friends' problems (4 weeks)

3.2.2 Materials for MPP Instructions

Ms. Wonder developed materials in two steps: (1) math content, and (2) MPP types.

Step 1. Math content.
The teacher chose four units on number sense. They were chosen because of their rich contexts (e.g., Moon Festivals) and their connections to each other (e.g., one-step to two-step problems).

1. One-step problem with a bracket: where a bracket is on a different position in a fill-in-blank expression
    e.g., $6 + (\ \ ) = 19$, or, $27–12 = (\ \ )$
    Two-step single operation problem by having one-step twice

e.g., 9–4 = 5 then 5–3 = 2.

2. Two-Step Mixed Operations Problem
   e.g., 9–4 = 5 then 5 + 2 = 7
   Finding multiples of or the total for how many '2's, '4's, '5's, or, '10's. Writing a chain of addition: 4 + 4 = 8; 8 + 4 = 12; 12 + 4 = 16; 16 + 4 = 20
3. Use tens and ones to solve addition and subtraction problems for numbers under 100. Finding multiples of or the total for how many '3's, '5's, '7's, or, '8's.
4. Solve addition and subtraction problems for numbers under 200; Find multiples of or the total for how many '9's, or, '12's.

Step 2. Types of MPP tasks

Three types of MPP tasks were used: Math Expression, Text, Drawn. Wonder made attractive worksheets to go with the MPP activity. Below are sample tasks and children's responses from a pilot study. (*Children's work will be typed using another font in a larger size and in italics.*)

Sample task-1: Math Expression.

Dear Children: Ms. Wonder wants to use the expression 35 + ( ) = 48 to pose a math problem to test the children in the class. However, she has a cold and headache and cannot pose problems. Please help Ms. Wonder to pose a math problem. ^0^ Thank you so much.

The problem I help Ms. Wonder to pose is:

There were 35 guards in the palace. The King hired some guards. There were 48 guards then. How many guards did the King hired.

Sample task-2: Text

On the day of the winter solstice, mother cooked delicious glutinous rice balls! There are 26 red bean glutinous rice balls, 17 peanut glutinous rice balls, 32 sesame glutinous rice balls, and 15 fresh meat glutinous rice balls. Everyone was happy to eat hot glutinous rice balls.

The problem I like to pose is:

Sandy ate 10 red bean and 5 sesame. How many did she eat?

Sample Task-3: Drawn

Milk \$62 Coffee \$25 Yogurt \$38 Water \$17

According to the above figure, the problem I pose is:

Father loved to drink fresh milk. He went to buy 5 bottles of milk. He drank 3 bottles. How many bottles of milk were left?

I can solve the problem.

$5 - 3 = 2$

3.3 *Data Collection and Analysis*

Data collected were children's posed problems (MPP), problem-solving test scores, children's feedback sheet, and teacher's journal. For oral posing, the teacher audio-taped and then recorded responses in written form.

3.3.1 For MPP, Children Posed While They Were Having Math Lessons (First Two Units), and Then, for a Change, They Posed Problems After They Had Math Lessons (The Last Two Units)

Coding MPP responses. Wonder read about the five types of categorization schemes by Leung and Silver (1997) and found that nothing was mentioned about mistakes in language. Ms. Wonder taught both language and math; she decided to evaluate students' outcomes on both. After piloting and having meetings with her peers and advisor on the study she decided on a 9-point categorization, as below. In the evaluation, she also considered language use such as typos, missing words, and, poor sentences (see Table 1).

Wonder's coding scheme for MPP responses (extended from Leung & Silver, 1997)

Table 1 Wonder's 9-point categorization coding scheme and examples

Code		Remark
9	Solvable math problems with complete structure	A candy cost \$3 and I bought 12 candies. How much did it cost in total?
8	Solvable math problems with complete structure but including typos or missing words	I had 18 pencils. Mom had me 14 pencils. I had 32 pencils. Father gave me 9 pencils. How many do I have then? *Note: "pencils" is missing*
7	Solvable math problems with complete structure but including poor sentences	My elder sister buys a watch that costs \$120, and my elder brother buys a book that costs \$45. Add sister to brother to get how many in total? *Note: add expenditure, not persons*
6	Solvable math problems but without connection to given information	Worksheet: 102–6 = ( ) The elder brother had 102 candies. Younger brother ate some. 6 candies left? *Note: Mistook as 102-( ) = 6*
5	Unsolvable math problems with insufficient conditions or inconsistent with the given	Mom took me to buy fruits, mom bought apples, grapes, and longans. How much did Mom spend in total? *Note: Cannot solve, insufficient conditions*
4	Unsolvable math problems and unclear	The elder sister had 42 fish first. Some fish died. Were there 26 fish altogether?
3	Non-math problems	Peter has \$100, he bought a toy car for \$28. What is the color of this toy car?
2	Just narrative	There were 35 guards in the palace. The King hired some guards
1	Incomplete narrative	The elder brother has 47 candies, and the little brother has

3.3.2 Problem Solving Tests (20 Items: 8 Text, 8 Expression, 4 Drawn)

As in problem posing, the math contents for solving were also on the four selected chapters. The test items were piloted to other students; and then asked for checking from Professor Leung, colleagues, and graduate students. Ms. Wonder selected items that were of medium-level difficulty in the final tests. There was a split into two tests (one for the pre-test and the rest for the post-test). There were two types of items. Drawn items: students wrote a math expression to solve the problem. Text items: students chose from four choices of math expressions after solving.

Sample Drawn Item

( ) biscuits

Sample Text Items (with four choices)

I wrote 64 characters today. My brother wrote 21 more characters than me. How many words did my brother write today?
(1)21-( )=64 (2)64+( )=21 (3)64-21=( ) (4)64+21=( )

For problem-solving test scores, Wonder did a t-test to compare the performance of her class and the class as a comparative group (did not use MPP).

3.3.3 Feedback Sheets

Dear children, Ms. Wonder found that there were a lot of good problem solvers in our class! The problems they posed were all very great, cute, and creative. Do you feel the same way? Please answer the following questions according to your own opinions and let Ms. Wonder know what you think!

1. Do you like to pose a problem? Why?
   □ I like to pose problems, because
   ____
   □ I don't like to pose a problem, because
   ____

2. Do you think it is easy to pose a problem by yourself? Why?
   □ I think it is very easy to pose a problem by myself, because
   ____
   □ I think it is very difficult to pose a problem by myself, because
   ____

3. Which is easier for you, and why?
   □Math expression is easier for me, because

____

☐Word expression is easier for me, because

____

☐Drawn is easier for me, because

____

4. Could you solve most of the problems posed by you?
   ☐Yes
   ☐No, because______________________

5. Could you solve most of the problems posed by classmates?
   ☐Yes
   ☐No, because______________________

6. Did you listen attentively when your classmates shared the problem they posed?
   ☐Yes
   ☐No, because______________________

7. Which is easier to pose a problem? "Orally" or use "writing"?
   ☐It is easier to use "orally"
   ☐It is easier to use "writing"

Children's feedback sheets were analyzed qualitatively to evaluate the MPP program.

4 Results

4.1 Development and Evaluation of an MPP Program

4.1.1 Wonder's Four-Stage MPP Program (Grade 2)

Stage I. Oral Posing (2 weeks)
The first two weeks were the most challenging. Since children did not get used to MPP Ms. Wonder used an oral posing approach. She carefully thought of steps in conducting the lessons. Ms. Wonder presented information. She wrote an expression (or a text) on the blackboard or pinned a poster with drawings. The whole class paid attention to the teacher's presentation on a chalkboard. She invited children to talk through what they found from the given information. Finally, she said, "What problems can we make?" In using "we" (instead of "you") it meant that the teacher was one of the members who made up problems. "Doing things together is the warmest

motivation." Ms. Wonder reminded herself constantly. In addition, she constantly repeated what children said and helped those who were too shy to speak.

Note.

Stage I helped students to get used to this activity. Ms. Wonder made a lot of effort to watch and constantly gave reminders. For example, a student made up a problem by setting the unknown but forgot to ask a question. An example is, "Chun had $16, and Mom gave her some money. She had $25 then." (Unknown Change).

Ms. Wonder reminded the students that a question was needed to solve a problem. Professor Leung consoled Wonder. Even though there was a reminder, young children found difficulty in seeing the difference between a description and a problem. Finally, Wonder added a question when children still forgot.

Wonder used Oral Posing (Stage I) to get students to experience all three types of MPP tasks. Students expressed freely what problems they thought of. Through oral posing, they learned to stay quiet when they listened to friends. "Appreciate the work of others and pay respect (MOE, 2008). They clapped hands when they liked friends' problems. Ms. Wonder also showed them how to nicely report when they spot errors. After Stage I, Ms. Wonder decided to let children try written posing.

Stage II. Written Posing (no need to solve) (3 weeks)
The same as in Stage I but requiring written work. Ms. Wonder normally made very attractive worksheets. She was experienced and learned that after students got the sheets they might not listen to her anymore. Therefore, Ms. Wonder explained before distributing the worksheets to excited students. When she wanted to focus on individual posing, she asked them not to talk with friends yet. She also told them not to solve the problem yet.

The teacher collected work and used her scoring rubric to grade them.

She used her scoring to design the next lesson. She returned worksheets. During the whole class, she shared posed problems and discussed selected work. The follow-up activity included appreciating or revising. For those who needed special attention, she did individual interviews later (Table 2).

Table 2 Some example materials for written posing (Stage II)

Single step	Two step	Adding several "6"s
Unknown Change	Mixed operations	Adding "4"s accumulatively
47 + 25 = ( )	18 + 14 = 32	6 + 6 + 6 + 6 + 6 = ( )
27–12 = ( )	32 + 9 = 41	Adding five "6"s all at once
102–6 = ( )	One type of operation	4 + 4 = 8
Unknown result	36 + 20 = 56	8 + 4 = 12
35 + ( ) = 48	56–18 = 38	12 + 4 = 16
42-( ) = 26		16 + 4 = 20

Note.
1. An MPP task should not contain many pieces of information
2. For the MPP type with math expression, Ms. Wonder checked if the problems matched the real life (e.g. the reasonable amount of rice balls per meal)
3. Ask students to check after writing. Ms. Wonder allowed ample time to correct written work such as wrong or missing characters and words

Stage III. Pose then solve your problem (3 weeks)

Attention:

There were two types of "pose-then-solve-own problems".

Type 1: Did not know they have to solve the problem posed by themselves (During Stage II)

Ms. Wonder carefully selected work from Stage II, sent it back to the student and this student solved that posed problem.

Type 2: Knew they had to solve their own problem posed by themselves and there was no need to send it to the teacher.

Note.

Stage III helped the teacher to follow a student's work and arrange for individual guidance in instruction. For example, a student made a mistake in solving. Ms Wonder needed to check if the mistake was in posing, or, in solving.

Example of mistake needed checking:

> "*Tai had $72 and bought a coffee using $25 and a bottle of water using $17, how much did he spend altogether?*"

Solution:

72–25 = 47,
47–17 = 30

This student did not find the total amount spent; he found the amount of money left instead. The amount of money he spent altogether should be $25 + 17 = 42$. Ms. Wonder interviewed this student to check if he did not pose properly what he intended (find the remaining amount) or if he posed properly but he did not understand the problem and solved it incorrectly. After an interview, she found that the students made mistakes in solving.

When students learned that they were to solve they might adjust the level of difficulty level of the posed problem. Most of the students understood the problems they posed, and then solved problem them without difficulty. When they were more confident, they posed harder problems.

Unlike prior stages, students read and checked their work during stage III, especially when they knew they needed to solve their own problem (type 2 of stage 3).

Stage IV. Pose and solve friends' problems

During Stage IV, students posed and solved problems. However, they exchanged problems they made freely or invited the teacher to randomly assign problems to other students. Then they solved the problems that were posed by their friends. This time, students needed to clarify, discuss, and work on MPP with friends.

As above, the design of this program was completed after several initial try-outs with tasks.

4.1.2 Evaluation of MPP Program

From regular meetings. Ms. Wonder and Professor Leung met weekly over the 12 weeks and also after the teaching. They evaluated the MPP program, the design, and the enactment of tasks. Ms. Wonder designed this MPP program as if she were a tailor (Leung, 1991). In Stage I, they learned to pose by orally presenting ideas. From Stage I to Stage II, children wrote what they spoke. As they tuned into Stage III, they posed and solved problems posed by themselves. However, Ms. Wonder was afraid that children posed a problem they could not solve. So, she read posed problems from Stage II and carefully selected one problem for each student. Type I was special in that Ms. Wonder did not choose problematic problems. As time went by and when the teacher made sure that students were more used to solving Ms. Wonder attempted pose-then-solve type 2. This time, children knew beforehand that they were to solve right after they posed problems. Stage IV, the last stage, was the hallmark of Ms. Wonder's program. During this stage, trading problems to solve was very exciting. They learned to read friends' problems, which was challenging. All in all, the steps were successful for this age group, and a total of 12 weeks were needed for the whole transition.

From Feedback sheets.

1. Do you like to pose a problem? Why?
   35 out of 36 said LIKE. The one who did not think so said it took S too much time to think of a problem. The 35 students expressed their feelings about LIKE. Up to 23 said they LIKE because MPP was *"interesting", "fun", and "creative"* and expressed that they were very happy doing MPP. Seven of them thought MPP was simple and they could write very quickly. "*When I read the information, I can make up a problem automatically*". Four of them considered MPP as "*tough training*". They were trained to be math elite, brilliant, and, brain-storming information once they read MPP tasks. They learned a lot. Finally, one student said, "*I like math and all activities during math lessons, including MPP*".
2. Do you think it is easy to pose a problem by yourself? Why?
   A total of 30 said it was very easy and 6 did not think so.
3. Which is easier for you, and why?
   Math expression-20
   Text is easier-7
   Drawn is easier-9
4. Could you solve most of the problems posed by you?
   All. This meant when students posed, they understood. When they understood they could solve problems.
5. Could you solve most of the problems posed by classmates? 35 said "yes" and one did not think so.
6. Did you listen attentively when your classmates shared the problem they posed?
   34 students said they loved to listen to friends' ideas and were learning to stay quiet when friends presented work. They also learned to share ideas. Sharing and accepting others was a goal for the curriculum standards.

7. Which is easier to pose a problem? "Orally" or use "writing"?
   It is easier to use "orally" (n=21). It is easier to use "writing" (n=15).

4.2 *Analysis of Students' Outcomes in MPP*

4.2.1 Students' Outcomes in MPP from Four Stages

STAGE I-Oral MPP
Ms. Wonder helped by reading out loud or repeating what the child said if the voice was soft.

Sandy ate 10 red beans and 5 sesame. How much did she eat?

The whole class clapped hands.

Ms. Wonder then recorded (in her notebook) the results of oral posing.

STAGE II-Written MPP
Harry Potter wants to challenge you with math problems to see if your math mind is as good as his mind. You are smart. Please accept the challenge of Harry Potter!

The expression given by Harry Potter is: $6 + 6 + 6 + 6 + 6 = (\ \ )$

The problem I pose is:

Mom went to the traditional market this morning. There were 6 eggs in a box. Mom bought 5 boxes of eggs. How many eggs did Mom buy?

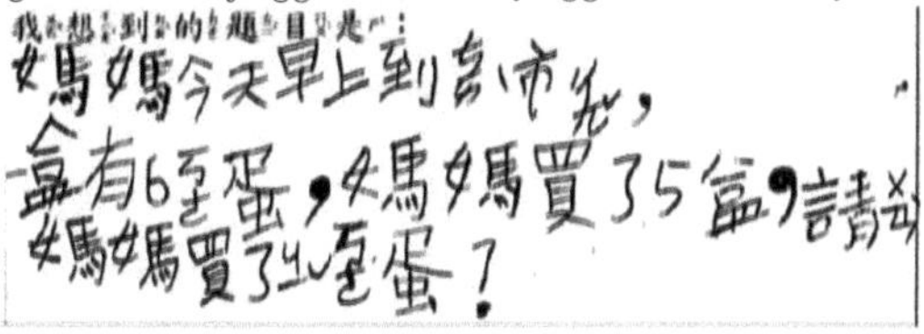
我想到的題目是：
媽媽今天早上到[illegible]市[illegible]，
一盒有6[illegible]蛋，媽媽買了5盒，請問
媽媽買了幾[illegible]蛋？

STAGE III-Pose and solve your own problem (3 weeks)
There were many animals in Mr. Wang's farm! He counted very carefully and found that there were 48 roosters, 35 hens, 19 goats, 24 sheep, 27 cows, and 13 pigs on his farm. You are smart. Can you use the above information to pose a great problem?

According to the above information, the problem I pose is:

Mr. Wang has 24 sheep and 35 hens. How many more hens than sheep?

I can solve the problem:

$35–24 = 11$

根據上面的資料，我想到的題目是：
王老先生有24隻綿羊，35隻母雞，母雞比綿羊多了幾隻？
我會把答案算出來：
35 − 24 = 11

STAGE IV-Pose and solve friends' problems (4 weeks)
During Chinese New Year, everyone received a lot of red envelopes (lucky money)! Junjun got $150, Mei got $187, Xiang got $106, Dahua got $88, Jia got $90, Ming got $120. They wanted to use the lucky money to buy snacks

According to the above information, the problem I pose is:
Mei had $187. She spent some money to buy pudding and $168 was left. How much was the pudding?
I ask (Lin) to solve
187–19 = 168
187–168 = 19
☑ He is right
☐ He is wrong. I ask him to solve it again

根據上面的資料，我想到的題目是：
美美有187元，買ㄆㄨˋ丁花了一些ㄑㄧㄢˊ，還ㄕㄥˋ下168元，請問ㄆㄨˋ丁多少元？
我請（林佑相）來算算看
187 − (19) = 168
187 − 168 = 19
☑ 他算對了。
☐ 他算錯了，我請他再算一次。

4.2.2 Analysis of Students' Problematic Posed Problems

Ms. Wonder found problems with math or language and prepared to interview students who posed them.

1. MATH

M1. Example of "Inconsistent with the information provided":

The expression given by the teacher is 27-12 = ( ).

"There were 12 fish in the fish tank, and the mother put 27 more fish. How many fish were there in total in the fish tank?"

Note:

The problem posed by the student did not match the numbers provided by the teacher. Ms. Wonder checked if the student (a) misread the expression to be 27+12 = ( ) or, (b) had problems understanding what the expression represented (not a reading problem).

M2. Examples of "do not match the fact in real life"

A pig has 5 legs how many legs do 4 pigs have?
Pigs only have 4 legs. The problem does not match the facts.

M3. Examples of "insufficient data":

Preston was 186 cm tall. We didn't know how many cm he had grown. How many cm he had grown?
There is not enough information to find how many cm Preston has grown.

There were two that did not reach the requirement on math problems, although there was no error.

M4. Example of "obvious":

The bird picked fruit and 12 fruits were eaten by birds. How many fruits did the bird eat?
The answer is already given. It was not a math problem that needed to be solved. It is an example of "obvious".

M5. Examples of "narration only":

There were 47 frogs in the pond and 25 more frogs came later.
No problem was asked, only narratives.

2. LANGUAGE
   L1. Example of a "typo":

Today, Dad went to the stationery store and **sold** *27 packs of paper, and used up 12 packs. Now how many packs of paper [Bowie's note: not sure if "dice" should be "paper"; otherwise, it should be also included as an example of inconsistent objects.] does dad have left?*
This student wrote the wrong word "**sold**" rather than "bought". The opposite meaning will affect choice the choice of addition/subtraction.

L2. Example of "missing words":

Mom went to the supermarket today to buy ( ) cost $62 and yogurt cost $38. How much did she spend in total?
The item that costs $62 is missing.

L3. Example of "missing units":

There were 27 loaves in the bakery. 12 loaves were bought by two customers. How many ( ) were left now?
"loaves of bread" should filled in (). The lack of the unit confused students.

L4. Examples of "confusing units":

There are 9 pens in a box of pencil cases. How many pens are there in 4 boxes?
Note: Box, pens, and pencil case are different and the units made the problem confusing.

L5. Examples of "inconsistent units":

There are 2 lenses in a pair of glasses. How many lenses are there in 5 glasses?
Originally the quantifier was "a pair of glasses", but later was "glasses".

L6. Examples of "inconsistent objects":

Mom had 14 strawberries. After eating a few, there were 10 strawberries left. How many candies did Mom eat?
Originally, the problem asked about strawberries but later asked about candy. One cannot solve this inconsistent problem.

L7. Example of a "poor sentence":

The elder sister bought a watch cost $120, and the elder brother bought a book cost $45. How much did they add up?
It should be adding the amount each person spent and not adding the people.

L8. This is an example of a "semantic error".

Maruko originally had 27 toys. She sold 12 toys at the flea market. How many toys does Maruko have in total?
The solution of the topic cannot be matched with the problem description. The problem solver expected to find the number of toys left.

4.3 Difference in MPP and Problem Solving

Here, the investigators wondered if there was a difference in MPP among three types of tasks and two kinds of subjects (task and subject). The two classes (with and without MPP) were also compared using scores in the problem-solving test.

4.3.1 Three Types of Tasks

Math expression (mean = 7.71), Text (mean = 8.11), Drawn (mean = 8.09)

Finally, these three types are used for dependent sample T-tests in pairs to compare the differences between them. It was found that students did MPP in Text and Drawn formats, and they performed better than in math expression. Children in grade two did not come across math expressions so often and they were not familiar with symbols yet. Students were better able to pose problems from pictures and written contexts than from formulas, but there was no difference between pictures and written contexts.

4.3.2 Two Kinds of Subjects

According to the scores of the students' MPP, the top 27% belong to the group with high ability in posing; the bottom 27% belong to the group with low ability in posing. It was found that the performances of the high problem-posing ability group and the low problem-posing ability group had a statistical difference. The high problem-posing ability group performs better in speed and correctness in solving problems than the low problem-posing ability group.

4.3.3 Relationship of MPP to Problem-Solving Performance

Ms. Wonder used self-developed tests on problem-solving in the number sense on 4 target units. She used a t-test to compare the test scores of the two classes. Her class performed better than the class (comparative group) which did not use MPP in teaching. Problem-posing teaching techniques appeared to help students to enhance their problem-solving ability. There was also an overall improvement in doing MPP after four stages in Wonder's program on MPP.

5 Conclusion

Wonder responded to the call for teachers to be curriculum developers and for students to take charge of their learning by actively posing a problem to solve (MOE, 2008). Similar to prior attempts (Skinner, 1991; Tsubota, 1987), the investigators concluded that even young children can do MPP. The condition is a careful program (in this case, four sequential stages) and a patient class teacher who knows children very well. Under Ms. Wonder's design, MPP and MPS were tightly integrated, and under this rubric for coding, math and language requirements were emphasized. Wonder extended the categorization of students' outcomes (Leung & Silver, 1997) to include not only math but also language accuracy, which was found to be useful for young children who were at the early stage of trying to read and write. Through this attempt, Wonder learned more about integrating subject domains (e.g. language use and math subjects), which was also advised in the curriculum document (MOE, 2008). Presentation of posed problems for appreciation and critique was also emphasized. The values of this MPP program also included active participation in discussion. No matter whether orally or in written forms, children's communicative power increased while creating problems. All these matched what was advocated in the curriculum guidelines (MOE, 2008).

Designed tasks should not give too many details, because, the more restrictions the less creative in MPP. For tasks with Drawn and Text format, there might be extraneous pieces of information. The teacher needed to remind students to write explicitly the numerical information content for solvers. For tasks with Expressions like "15 + 7 = ( )" there was no need to do so. Finally, she was thinking of linking MPP from school to home (Leung & Wu, 2000). With help, children posed problems with their parents at home and were able to do revisions. Students could be reviewers of problems so that they better understand the posed problem given by friends.

References

Carpenter, T. P., Fennema, E., & Franke, M. L. (1996). Cognitively guided instruction: A knowledge base for reform in primary mathematics instruction. *The Elementary School Journal, 97*(1), 3–20.

English, L. D. (1998). Children's problem-posing within formal and informal contexts. *Journal for Research in Mathematics Education, 29*(1), 83–106.

Heller, J. I., & Greeno, J. G. (1978). *Semantic processing in arithmetic word problem-solving*. Paper presented at the Midwestern Psychological Association Convention, Chicago.

Leung, S. S. (1991). Building connections to the Pythagorean Theorem: An example of teachers' treatment of textbook problems. In M. K. Heid & G. W. Blume (Eds.). *Mathematics Education: Making connections*, pp. 41–47. Pennsylvania Council for the Teachers of Mathematics.

Leung, S. S. (2013). Teachers implementing mathematical problem posing in the classroom: challenges and strategies. *Educational Studies in Mathematics, 83*(1), 103–116.

Leung, S. K., & Silver, E. A. (1997). The role of task format, mathematics knowledge, and creative thinking on the arithmetic problem posing of prospective elementary school teachers. *Mathematics Education Research Journal, 9*(1), 5–24. Australia.

Leung, S. S., & Wu, R. (2000). Sharing problem posing at home through diary writing: Case of grade one. *The Australian Mathematics Primary Classroom, 5*(1), 28–32.

Ministry of Education of Taiwan. (2008). *General guidelines of grades 1–9 curriculum for elementary and junior high school education* (in Chinese).

Skinner, P. (1991). *What's your problem: Posing and solving mathematical problems. K-2*. Heinemann.

Tsubota, E. (1987). *On children problem posing* (Grade 1 to Grade 3). Guotu She. In Japanese.

Chapter 10
Surfing Over Three Waves in Mathematics Problem-Posing Teaching for Third Grade: The Voyage of Anna and Examples on Mensuration and Shapes

1 Introduction

The importance of math problem posing can be inferred when curriculum standards advocate teachers to provide opportunities for children to do MPP (MOE, 2008, 2018; NCTM, 1991). Taiwan curriculum standards also suggested teacher to do action research. A grade three class teacher, Anna, attempted to integrate math problem posing (MPP) and be an action researcher. In paying attention to MPP, she aimed to study how to integrate, the possible difficulties, and how to overcome them. She also believed that attending only to problem-solving was insufficient because it would only tell what children did not know but not what they did know (Leung, 2013). If the attention was put on MPP we could study how children posed problems and judge what math they know about a situation. In studies on MPP, the investigator often attended to children's behavior instead of teachers' attempts to teach. In all, Anna the investigator her co-investigators (her advisor, Professor Leung), reported on professional development through this attempt.

The purposes of this study were three. First, Anna designed a program and walked through an action research journey. She captured what an implementation of math problem posing in elementary school (grade 3) would look like. Second, she recorded the possible outcomes given by students (grade 3) and delimited the report to include examples of time, length, and angle. Third, she reflected upon practice and closed with her reflections on this voyage.

S. S. Leung, *Mathematics Problem Posing in Action*,
https://doi.org/10.1007/978-981-96-8480-9_10

2 Anna's Background Literature for Actions

2.1 *Teachers as Action Researchers*

In each curriculum change, there was also a need for teachers' change. Teacher educators and teachers were engaging in action research to implement related changes in teaching and learning emphasized in the curriculum change. They considered the kinds of knowledge a teacher draws upon trying to be a curriculum maker. Clarke (2008) told the story of Ms. X trying to teach fractions using what she learned during professional development sessions. This teacher reflected on what had happened in the previous related lessons. She then read curriculum guidelines and thoroughly read the ideas related to the lessons she taught. She also read literature related to her teaching (on fractions) to link practice to research findings. She learned from the papers to interview students and to bring her lesson structure. The research influenced her in making assessment tasks and in encouraging students to share the results of discussions among groups. Clarke (2008) referred to this story and traced the kinds of knowledge involved in the process of teaching by Ms. X. The teacher's knowledge included content knowledge, pedagogy knowledge, and pedagogical content knowledge (Shulman, 1986). Ball and colleagues (2008) elaborated Shulman's content knowledge and pedagogical content knowledge into six distinct domains. The six sub-categories are common content knowledge (CCK), specialized content knowledge (SCK), horizon content knowledge (HCK), knowledge of content and teaching (KCT), knowledge of content and students (KCS), and knowledge of content and curriculum (KCC) (Ball et al., 2008; Hill et al., 2008). Anna knew that she must equip herself with various kinds of knowledge to teach math: content and pedagogy. In terms of teachers as curriculum makers, knowledge of the curriculum is important. Especially when the change involves different sequencing of topics (what to teach) or the ways to teach children (how to teach).

2.2 *Mathematics Curriculum for Grade Three*

When teachers teach measures and shape (*what to teach*), how can they enact tasks that include problem-posing and problem-solving on the part of children (*how to teach*)? Anna first studied *what* and then attempted to explore *how*. For example, she used the following curriculum map to match the grade three curriculum guidelines (see MOE, 2008, 2018). Anna and Professor Leung referred to Fig. 1 for carrying out this investigation. They also read all topics in the grade three textbook series. There were two major strands.

For Mensuration, children experience amounts and quantities from pre-school. In elementary schools, children began to learn quantities, which was important in linking them to learning numbers (Smith & Barret, 2017). Smith and Barret include a discussion on five physical quantities prominent in students' experience, adding in school

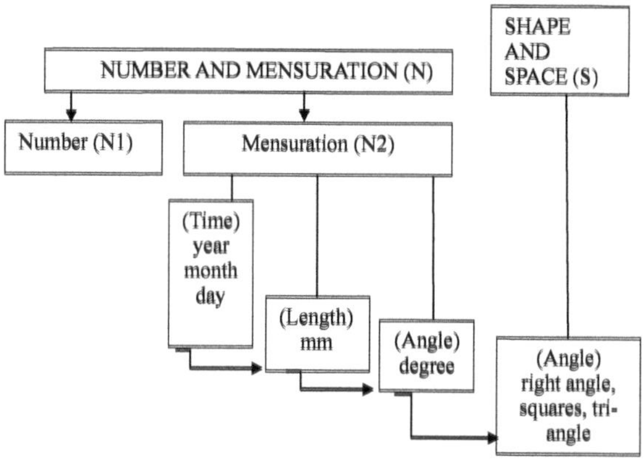

Fig. 1 Curriculum map for Anna's investigations (Grade three)

learning, and published research—length, area, volume, angle, and time. In Taiwan math curriculum standards (MOE, 2018), the above five (and three more) were given under Mensuration (N2, Fig. 1) across different grade levels. N2 is under one of four strands called Numbers and Mensuration (N), where N1 represents Numbers. The first three (length, area, volume) are related and differ in dimensions. The fourth, angle, is the measure in degrees and before the use of a protractor, a child can compare sizes of angles by overlapping. The last of the five, time, is abstract and would be concretized by the instrument that measured it (e.g. the sand dripper); children learn about Year, Month, and Day in grade three. The curriculum guidelines on mensuration for Taiwan elementary schools also include money, weight, and, speed, adding to eight types of measures. The content "angle" appears under N2 and S. On angle, it includes angular measure (N2, grade 3) and also angles in shape and space (S, for grade 3, see Fig. 1).

2.3 *Knowledge of Teaching for Mensuration*

Notes on instructions were given under the curriculum guidelines (MOE, 2008, 2018). Teachers usually provide manipulatives, and hands-on experiences before using instruments to do exact measurements. Comparing and ordering two measures follows, then learning to combine measures or decompose one. Finally, children learn the use of proper units to represent a quantity and conversion of units. The caution for teaching is to avoid over-emphasis on procedures and conventions. What is more important is rather on the development of students' understanding of the particular measure and its quantity. The part on procedures such as conversion of units, can be found in the learning of numbers and calculation (N1). For instance, children's learning of area concepts is more important than an exercise on the computation of two numbers for the area of a rectangle (multiplication, "base times height").

Before grade 3, students already learned some number facts and calculations (multi-digit numbers, addition, under N1) and measures (length in cm, under N2). By grade 3, they understand its unit (e.g. cm), then compare two measures (e.g. which is longer). Later, they get to know larger units (m), smaller units (mm), and conversions from larger units to smaller units or vice versa. Finally, they learn to use appropriate instruments to measure the size (e.g. finger span, ruler) and tell the size of a measure in arbitrary (a toothpick) or standard (a ruler with cm and mm graduation on it) units. For shape and space (S), Anna read the teachers' guide of textbooks.

As above, Anna equipped herself with an understanding of action research, curriculum knowledge, and knowledge for teaching. She was ready to Bon voyage with Professor Leung and surf through waves in MPP.

3 Method

By adopting the action research method, the research was conducted for a semester in the mathematics class of the third grade which the researcher had been teaching.

3.1 Participants

Grade three students. The sample consisted of a total of 40 children (18 boys and 22 girls). The school was geographically located in central Taiwan. Children came from families with parents' occupations mainly from farming and industry. Children were having the new curriculum from grade three that emphasized student-centered and group work. However, children who were taught in grade one and grade two using traditional ways; seldom voiced out own ideas interacted with the teacher, and, barely worked in small groups.

The investigator. Anna, the investigator, graduated from a teachers' college that prepared elementary school teachers. During her fifth year being an elementary school teacher, she was pursuing her graduate study and decided to do action research as her master's thesis, and also found answers to her questions about using mathematical problem posing (MPP) as a means of instruction. She wanted to be a practitioner who connected research and practice, identified integration challenges, and enhanced her reflection skills through professional development evaluation.

Anna met her advisor, Professor Leung when she attended graduate school. Anna was a young and devoted teacher who taught in an elementary school in central Taiwan. The school she taught was an hour away long from her advisor by car and she drove after work instead of studying at a nearby university. This was because she read Leung's work on math problem posing and she decided to learn about it through research practice. Anna attended talks and a local scholar also convinced her that teachers could also be researchers and that teachers' autonomy better knowledge in teaching. She took a course on qualitative study and the part on action research attracted her. Finally, her thesis advisor supported her decisions in MPP and action research.

3.2 Development of MPP Materials

Problem-posing materials design was made by referring to Leung (2013), Tsubota (1987), and Winograd (1990). As for the selection of units, she used units that were rich in context, real life, and simple enough to do problem posing. As for types of MPP tasks she decided to use four: Expression, figure, table, and solution (Leung, 2013). After reading the textbooks for the proposed study, the researcher designed the MPP teaching activities for seven selected units (see Table 1, for seven "v"s in the column for MPP). The units were rich in content, connected to real life, and a bit challenging. Anna prepared for class with the same conscientiousness, the same way as she did before. In this chapter, data analysis was taken from three chosen units: **Time, length, and Angles** (see Table 1, **bold**).

The MPP activity took the format of both individual and group. Anna used the time wisely: Before teaching, during teaching, and after teaching. The teaching involved asking children to pose problems and the posed problems were collected as

Table 1 Three reported units (**bold**) among seven MPP units ("v")

Unit	Name	Contents	MPP
1.	Number less than 2000	Number sense, conversion, place values	
2.	**Time**	**Year, Month, Day**	v
3.	Division-1	Meaning Basic division	
4.	**Lengths**	**Using the rule with mm as graduation marks, measuring in mm**	v
5.	3 digit-number (+, -)	Record of results Vertical operation	v
6.	**Angles and shapes**	**Measure of an angle, right angles, squares, triangles**	v
7.	Weight		
8.	Multiplication	3-digit times one digit	v
9.	Fractions	Fraction, numerator, denominator less than 10, denominator less than 20	v
10.	Area and perimeter	Indirect comparison of area, individual unit comparison, and measuring areas, area, and perimeter	
11.	Division-2	Division with remainder, long division	v
12.	Bar Chart	Tally, data collection, draw, and read bar charts	

data. For those outcomes from "before teaching" the teachers used them as start-up tasks: presentation of problems to solve. For those outcomes from "during teaching" the teachers used them for teaching. For those outcomes from "after teaching" the teachers used them for homework or revision exercises or pin them up on notice boards for children to visit and admire at leisure or to do peer review.

3.3 Data Collection and Analysis

The teacher/researcher engaged in action research for six months. *Data collection* included videotaping, observation notes, interviews, reflection notes, and, activity sheets (posing, evaluation, and diary). Data was analyzed qualitatively using triangulation. *Data analyzes.* The qualitative data were collected and transcribed. The investigator read and analyzed according to her research questions. The resulting analyzed pieces fell into categories, themes, and patterns by using triangulation or constant comparative method. Anna was trying to find the best explanation for the data. She read the book by Lincoln and Guba (1985) to learn how to ensure *trustworthiness and limitations.* She reminded herself to try hard on data collection to increase trustworthiness and wrote limitations.

4 Results

Anna's belief. The action research on implementing curriculum standards depends on the investigator's belief. Anna learned about constructivism in the third year of her college, which also included the teacher preparation program for elementary school teachers. She was very interested in this way of getting children to learn when compared to the mechanical way of asking children to follow "do's and "don'ts. She believed that knowledge should be constructed by the learner and not transmitted from the teacher to the learner. When she started to teach, she had a heart for her teaching and taught with a passion. However, when she was immersed into the school setting there were times when she had a feeling of losing strength, especially when she found the gap between theory and practice.

> *"How could I fill this gap?"*

She carried this question to her graduate study. Despite being stressed with time, she talked through her thoughts with peers or other graduate students and professors. She believed that talking could help. She realized that each teacher used their own beliefs to support teaching. To her, all teachers differed in beliefs and therefore they taught differently. She believed that processes in learning math were as important as products in learning. She believed that getting children to think and reflect on what students did could lead to meaningful learning. In the past, teachers solved problems for children to learn by imitation. Under the climate of curriculum change that emphasized constructivism, the teacher only presented problems but did not solve them. It was the children who solved problems. Under this change, teachers converted the traditional "preaching, teaching, and solving" to the presentation of contexts and guidance on children's solving. Since knowledge was actively constructed from the individual's heart, not given by the outside world, the teacher changed the traditional role of the authoritative person in the past and became an accompanying adult who witnessed students' learning.

Anna believed that math classes could be made very interesting by having students' lives as part of the math classes. "*Let us do math problem posing and be very active.*" She believed that the classroom could be a paradise for math when children actively did mathematics. Anna considered play as a way to make children like math and math classes would not be nightmares. In addition, children learn to communicate their ideas and share them with others. Learning was a cooperative and discovery journey and not a lonely adventure. Children not only learn math but also experience ways of doing social interaction through play.

Anna is an action researcher. She read papers on problem posing in graduate study and she believed that this way of teaching matched her beliefs. Upon attending workshops on action research, she found that many school teachers were doing action research at that time. She was also attracted by the work of Silver who worked with teachers and developed cases as exemplars (e.g., Smith et al., 2005). Anna attended a talk given by Silver when he visited Taiwan; the talk was on cases as exemplars. She noticed that professors and teachers can work together to link research into practice.

With Anna's belief and determination to do action research, she followed the teaching sequence suggested in the curriculum guidelines and completed her study. First, she implemented MPP into her teaching accordingly and attempted to resolve teaching problems. Second, she recorded and analyzed students' outcomes. Third, in looking back at her implementation, she found traces of footprints along three stages in her professional development journey.

4.1 The Three Stages: Groping, Transition, and Maturation

Anna found herself surfing over three waves as she related to Professor Leung about her actions. She found herself groping around, tuning in, and getting on trek, during her voyage in teaching math using MPP. She taught a total of seven units using MPP.

1. *GROPING: The difficult start.* At the start, neither Anna nor the children were experienced in MPP. She originally planned to divide the class into collaborative groups of five. However, Professor Leung wondered if it was too risky to pose in groups of five, especially during the first unit. After a meeting with Professor Leung, Anna changed the five-person grouping to a two-person grouping (*0906 journals*). She wrote in her journal to remind herself to wait until the students gradually got used to this activity, then they would learn in groups of five (*0913 journals*). The two-person grouping applied to the second and third units (*0920 journals*). Before

teaching, she believed that her thorough preparation would enable students to pose wonderful math problems, rich in imagination, and incorporate things in life or popular cartoon characters into it. However, this difficult start led to a disappointing ending. Students' problems were plain and similar to friends (*0927 journals*). Anna desperately needed to change her way of teaching: How could she teach to help children pose problems from plain to creative?

2. *TRANSITION: Anna's attempts to re-solve teaching problems*
Below are teaching problems: *Lack of creativity, blank sheets, students not knowing how to discuss, and contradictions.*

Lack of creativity. During stage I (Groping) she found that students were not creative in posing problems. After she gave a deep thought and did a serious review, Anna thought that the students might be affected by the setup. She made attempts to solve this problem. *Attempt I—The magic bag.* Therefore, for the second activity on problem posing, the teacher specially carried with her a fancy magic bag with many interesting objects inside it. She hoped that the students would use the objects in the bag to pose a problem (e.g. about the lengths of objects inside the bag). Anna asked herself, "Will they all ask the same question: how long is XX?". As the teacher, she could remind students to think about other question types (*1005 journal*). In the actual teaching, it was as Anna expected. Students' problems were, "What is the length of X?" Anna was disappointed too at the objects students chose to use to make problems. They preferred to write about things they like to eat (e.g. bubble gum). Other things such as candles, straws, and batteries (*1006 journal*) were not mentioned. "*Dear students, you love bubble gums and ignore objects that you cannot eat.*" After this attempt (magic bag) she found that this teaching problem was not resolved: the posed problems were still lacking creativity. She needed advice. *Attempt II—Storytelling.* After discussing with Professor Leung, she changed the design of the next unit to story-telling and used stories to present the materials, hoping to improve students' creativity in posing problems (*1108 journal*).

"*One day, the Little Mermaid and her friends were going to the night market to buy shoes, but she saw Maruko's shoes and Little Red Riding Hood's shoes.*" This story was only a start-up. After she told this story, students were excited about the shoes. They were invited to be authors and continued to tell the story. As they continued to tell the story, they began to make up math problems. One student presented her work, "*She liked both pairs of shoes, so she asked the boss how much Maruko's shoes cost. The boss said it was $386. The little mermaid asked the boss how much was the price of Little Red Riding Hood shoes. The boss said that it was $100 cheaper than Maruko's shoes. How much was the price of the Little Red Riding Hood's shoes?*" (*1109 students posed 25*).

There was a transition period and indeed, after such a change in task setup using a magic bag, then using the teacher's story-telling and students' completion, the creativity of students improved (*1109* journal). Therefore, Anna wrote in her journal to remind herself that appropriate adjustments made according to the interests of students, made the lesson smooth.

Blank sheets. In a normal class, there would be several students with low academic achievement. Such students always need more attention and guidance from teachers. The class in this study was no exception. In the first unit, a total of four did not complete worksheets. Because of this, the teacher decided to guide these students. Therefore, when the students were brainstorming, she specially guided these students to help them find key points (*1109 interview*). "*If you help them to identify key features then they posed smoothly.*" *Attempt. Anna's intervention.* After the teacher's intervention, her students' work improved. Anna reported proudly during one meeting, "*Let's take a look at the work of S12 who always handed in blank sheets before I gave guidance.*"

> "*One day, Mr. Swallow was going to the Ancient Stationery Shop and saw Yongqi. Swallow walked over and asked Yongqi to accompany him to buy a $130 pencil case and $780 beads. How much did these two add up?" (S12 posed).*

Students not knowing how to discuss. To Ann's surprise, her third-grade students did not know how to discuss with regard to math problem posing. In the beginning, students' discussion was on superficial features such as poor handwriting or wrong characters. However, the teacher preferred them to talk about whether the written work was a math problem, check if the math problem structure was correct, or, talk about whether the problems were solvable. *Attempt-- Anna's intervention.* One example was the posing of a riddle or a poem but not math problems. Anna led the class to discuss the math part of the posed problems. Her solution to this teaching problem was not to mention the point by herself. The teacher did not comment but asked the children to read and comment. It was exciting. The students realized that those were not problems and asked the authors of these problems to pose again. She found reminding children to read again and comment worked out fine.

There were times when students did not know how to follow a discussion. The following is an example. When student S30 was talking about books, a student (S34) joined in but asked about robots.

> T: Come on, S30.
>
> S30: He didn't write the price of the storybook, and, with a proper unit called "yuan" (Yuan means $ in Taiwan).
>
> T: Thanks, S30... he didn't write "yuan"? Does anyone have any comment? S34.
>
> S34: Ms.! He should write four robots.
>
> T: Are robots related to the price of the storybooks? Why "four robots?"

At the time when someone (S30) sent a query about the missing price of a storybook (*1126, observation 159*) and that the problem had insufficient data for solving S34 added that S30 should write four robots (1126, observation 161). Anna found that S34 said something (four robots) that had no connection to the discussion on the price of a storybook. Anna directed S34 to follow what S30 said before, the price of the storybook, and to follow well when S34 joined in the discussion.

Contradictions. Many problems (posed by students) were not checked by students and contained contradictions. Even with reminders to check, the situation was the same. *Attempt—Read and do peer review.* Anna handled student contradictions by peer review. Peer review was a strategy Anna used to let students look at each other's work, appreciate others, and give their own opinions. However, for third-grade students, the teacher should lead the class to do group reviews before doing individual reviews. During the first review, Anna found that the students' answers were contradictory. For example, student S3 ticked off "*a very good problem*", but the reason was "*too many typos*". After she encountered this difficulty, the teacher reflected and let students practice beforehand. Hence, before the second review, she wrote one student's problem on the board and led a class discussion on each problem. Students spotted contradictory comments, voiced, exchanged ideas, revised comments, and made a consensus. Fortunately, results were satisfactory during the third and subsequent reviews.

3. *MATURATION: Anna's harvest time*

Meaningful learning and solving correctly. During this stage, Anna almost resolved most of her teaching problems and students got used to MPP. Problem-posing activities have made math learning more meaningful. Students felt in control of their mathematical learning through problem posing activities. Through problem posing activities, students made mathematical abstractions concrete and beneficial to students' mathematical learning (Borba, 1994). There were students who failed to solve the textbook problem but were able to correctly solve the problem they posed. This was because when the problem was posed by oneself; the problem could be easily understood and then solved.

An example (Min) could be used to explain that the problem posing activity can make students feel that the problem was simpler and the learning was more meaningful. Min initially failed to solve the textbook problem. It was from the third unit (Activity 5: Subtraction by vertical computation). Since the material for the problem is a structured problem, the problem posing activity requires students to first solve the problem and then imitate it to pose a problem related to addition and subtraction.

Solving textbook problems: There were 630 cows on the pasture and, the number of horses was 165 less than the number of cows, how many horses were there?

<u>Min's solution to cow-horse problem</u>

$630 + (\quad) = 795$

$$\begin{array}{r} 630 \\ +\,165 \\ \hline 795 \end{array}$$

Min's solution was incorrect. This subtraction should be solved by vertical computation. After posing the problem, Min's situation changed and he was able to solve his posed problem correctly.

Solving Min's posed problem: Fairy Moonlight went to a department store with her mother and saw a beautiful skirt that cost $220, and the price of the trousers was $150 more than the skirt, and the coat was $50 more expensive than the trousers, and her mother had \$600. If she bought a coat, how much money was left? (*1109 pose, S25*).

Min's solution to the skirt-trousers-coat problem

$220 + 150 = 370$

$370 + 50 = 420$

$600 - 420 = 180$

$$\begin{array}{r} 220 \\ +150 \\ \hline 370 \end{array} \quad \begin{array}{r} 370 \\ +\ \ 50 \\ \hline 420 \end{array} \quad \begin{array}{r} 600 \\ -420 \\ \hline 180 \end{array} \quad \text{(1109 pose, S25)}$$

Min's problem was a multi-step addition and subtraction problem, and the process of solving it was comparatively more complicated than the cow-horse problem given in the textbook. However, Min was able to solve the problem correctly. This example showed that Min made great progress. Since the problems were posed by the learner himself, the learner understood this problem very well. He was the solver and so, the problems could be easily solved.

Extending problem posing from school to home. According to Leung and Wu (2000), problem posing at home could make math more relevant to real-life applications, extend math learning from school to home, and promote interaction between teachers, parents, and students. In doing MPP at home, parents can share math ideas with their children. After five units of instruction using problem posing, students were asked to pose problems at home. It was found that posing problems at home increased parent–child bonding (*Interview data 2039, 2040, 2041, 2042, 2043, 2044*), which matched the idea of Leung and Wu (2000). Students and parents sent back written work of problems to Anna. She selected and used them as exercises during math classes.

As above, the action research cycles were re-visited by both Anna and Professor Leung. Each of them carried a meaning in Anna's stages of growth in this voyage on MPP. Professor Leung gave applause for her effort and witnessed Anna surfing through stages of groping, transition, and, maturation. Anna's extension to MPP at home hinted Professor Leung a new direction, to work with parents.

4.2 *Studying Students' Learning Outcomes: Time, Length, and Angle*

Anna arranged her tasks according to the instructional sequence given in grade three. Below, three units (two on mensuration and one on shapes and space) were used as examples to show how Anna studied students' outcomes and used the results in her teaching.

4.2.1 About Time

The Calendar (Days in a month, Sept. 15, video).

Children were invited to pose mathematics problems using a calendar (a month of the year). It was found that children were having different problem schemas and posed problems differently. When the class followed with a discussion the teacher had to decide on the order of presenting children's work. The investigator read and decided to sort them into types according to the instructional sequence of time in the textbook as suggested by the curriculum guide (days, week, month). According to her record, there were at least three types of problems: (1) Reading and telling what the date was, or telling how many Sundays in the month, (2) counting a certain number of days from a given date and telling the date, and, (3) calculate the total number of days between 2 given dates.

Anna showed students three examples, one from each type as above

1. *Which day of the week is Sept 25th? how many Fridays were there altogether?*
2. *Mom leaves home on Sept 5th. She returned after two weeks. Which day would she come back?*
3. *Dad was on a business trip from Sept 8th to 15th how many days was the trip?*

The above three examples showed what students know about time: Names of days in a week, the number of Fridays in that month, counting two weeks from a certain date and telling the date, and, counting the number of days between two given dates.

The teacher respected the students' work. Anna learned that the teacher must be alert and respond promptly when children pose problems relating to measures such as time. If the next step was to teach students to solve them, she arranged those posed problems from simple to complex. *"I can use them to teach after sorting and ordering the outcomes, and used them to replace the original textbook problems"* In doing so, children were excited to see that their problems were more important than problems given in the textbook.

Anna also needed to check if the problems were logical and followed real-life experiences. For example, children did not go to school on Saturdays and Sundays but one posed problem included only a holiday on the 22nd day. *Amy goes to school on Sept 1st and has a holiday 22nd how many days was she in school?* When the problems were problematic, Anna needed to guide the children to fix the problems. For example, the students needed to add the days when they were on weekends so that they could solve the problem.

4.2.2 About Lengths

The Unit "Millimeter" (Measuring in mm, Oct. 15 ~ , Video)

In this unit, children learned to experience length and how to measure. They already learned "cm" before and this time they learned a new unit called "mm". The teacher used both identical or different objects for the class to pose problems. She put objects into a magic bag and picked them at random. Children used the objects to pose problems. Anna sorted them into three categories: comparing lengths, connecting two lengths, or finding the difference between two lengths.

Comparing two lengths. (two identical square tiles) Compare the length of the sides. Which tile has a longer side? (answer: the same)

Connecting two lengths. The length of the box is 5 cm 1 mm and the length of the name card is 8 cm 5 mm. When put together, what are the total lengths?

Difference between two lengths. Chewing gum is 7 cm 4mm long and the ball pen is 7 cm 5 mm. Find the difference between the two lengths.

Again, Anna checked if the problems were logical and made sense. In recall, the teacher educator noted that the problem was legitimate for mathematics. However, in real life, connecting a box to a name card was not common. Also, it did not make any sense to compare the length of a piece of chewing gum to the length of a ball pen. Next time, Anna could try to suggest children make more practical problems: putting two-floor tiles together, or, comparing lengths of two pieces of wire for handicrafts.

4.2.3 About Angle (Angle Measure, Right Angles, Squares, and Triangles)

Angle (understanding shapes, Nov. 7~, video).

The task was about a figure with geometric shapes students were to use the figure to pose problems. Anna expected students to pose problems in space and shapes. She used a figure given in the unit from a textbook (Fig. 2) and invited her students to make up problems using the shapes given in the figure. There are twelve shapes, with triangles, and squares of different shapes and sizes.

The teacher asked questions to guide students to read the figure.

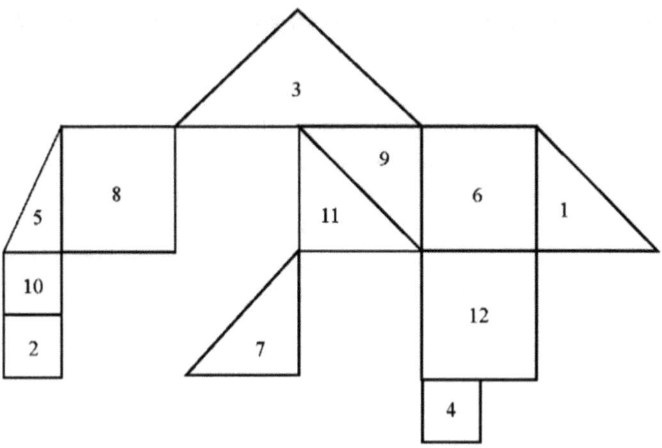

Fig. 2 Pose problems in space and shapes

T: I show you this picture, what do you see in this picture?

S8: I see triangles and squares.

T: Is there anything else?

S8: Vertex, side and angle

T: In this picture, what do you see?

S37: Triangle.

T: Triangle.

S37: Square.

T: Square, and then?

S37: Vertex.

After studying the figure thoroughly, they posed problems. Students posed problems and talked about features of the shapes (Vertex, side, and angle). They posed problems with the names of the shapes. Interestingly, they problems in number (N1), mensuration (N2), and shape and space (S).

Number (N1). *A square costs $10 and a triangle costs $11. I buy 9 squares and 1 triangle. How much should I pay altogether?*

Mensuration (N2). On lengths, "*Please measure the side of this figure in cm and mm.*" On angular measures, '*What is the angle sum of the figure marked "1"?*'

Shape and space (S). Features of the shapes, '*How many angles are there in figure number "4"? How many sides?*' Names of the shape, '*What kind of triangle is the shape number "3"?*'

In looking back, Anna said she should try to encourage children to make up problems even in data handling such as organizing data or finding the chance of picking one shape that is a square.

4.3 Anna's Reflection on How to Resolve Teaching Problems

By referring to how Anna reflected upon practice, the teacher educator witnessed Anna's growth in math knowledge for teaching (Ball et al., 2008). Similarly, traces of kinds of knowledge could be found in Ms. X, who learned how to teach fractions in Clarke (2008). Spotting problems in teaching and thinking about how to resolve them required pedagogical content knowledge. Below are examples of PCK (Shulman, 1986), divided into knowledge of content and students, knowledge of content and teaching, and, knowledge of content and curriculum (KCS, KCT and KCC, Ball et al., 2008).

4.3.1 Knowledge of Content and Students (KCS)

Anna learned about students' math abilities, and, how to manage the noisy class.

1. *Students' math abilities varied.* The teacher helped low-ability students avoid resulting in blank sheets in both posing and solving math problems. Individual interviews were arranged for extra guidance. When she found Fast Trek children starting to interfere with other children, she asked them to do related drawings so and would not get bored. For units where students failed to connect to real word problems (like an introduction to geometric figures), she used stories to attract creative MPPs. Finally, she modified the difficulty level of tasks and made them proportionate to the level of students (Polya, 1945).

2. *Noisy class.* Some teachers would consider noise annoying when children's voices were too loud. At first, Anna found the noisy class was a teaching problem. Later, Anna found that there were times in math lessons when making noise was not a problem because students were enjoying MPP and felt excited. For example, when two students gave different answers, both were correct (drawing two different shapes that match the same description for lines and angles). "*Yeah, this is correct*?" "*Well, the other answer is also correct*". The investigator then accepted this noise class, because this was because children were too excited or too serious in communicating ideas or making their contributions.

3. *Presentation of work.* She lost time when children were ineffective or shy (or slow) in writing when they came up to the board to present the results of class work. She learned to use small whiteboards and let them write in their seats. When she needed to share in a whole class discussion, there was no need to write again. They brought the small whiteboard and just pinned up their works to the blackboard. When she was indecisive about which pieces of work to be presented and discussed, she sorted after school and shared work the next day. It was the sorting after class that made her learn about how to teach the next day for that particular math topic. Gradually the presentation part was under control. In discussing work produced by peers, she directed children to pay attention to math contents (e.g., conversion of units in a year, month, and days for a leap year) rather than superficial features (like mistakes in copying the names of children in the problems).

4. *Peer review.* Anna found that reviewing had to be taught. It was hard for two by two to do peer review and often led to contradiction. So, she selected one piece of particular MPP work and did a whole class review together. Later, when the children were more experienced, she let them do it in groups of two. It worked! They did improve the quality of the review.

4.3.2 Anna Gained Knowledge of Content and Teaching (KCT)

Introducing MPP. First, she learned about class and set up the class daily routine. Before integrating MPP the teacher must get the class ready for daily routine and understand the children's mathematics abilities. At the beginning of the school year, she spent much time on classroom management and tried to get them ready by learning about this class. "*If I get them ready as soon as possible, then there is no need to lose time on minor issues and I can focus on teaching.*" To tune in smoothly, she found time to chat with the class teacher of the previous year and understand the whole class more to have a head start. She noted down the individual differences of her children. Second, she was flexible and edited tasks when needed. As the teacher and children first met in September both need to adjust to each other. When the design of activities did not work the teacher changed quickly to meet the needs of children. Third, she used the term "make a problem" instead of "problem posing" because it was easier for third-grade children to understand.

Set-up Tasks for MPP. When MPP tasks were set up properly (Butts, 1980) they attracted children to make up problems. When tasks were too difficult or too many or when Anna's instructional design did not match children's level of understanding she would check her choice of tasks or set up of tasks. After realizing this, the teacher modified tasks and made them simpler. When there was the dilemma of having too many activities, Anna would cut down the number of tasks. For example, the third unit consisted of too many parts and caused delays. The teacher decided to focus on posing and its discussion and skipped the solving part.

Problems arising. During stage one she had a hard time when displaying children's problem-posing products, she wasted a lot of time. When she began this kind of instruction children made up problems that were problematic and needed a lot of time to revise/rewrite. In addition, she also spent a lot of time deciding on the standards and format of problems to be displayed. How to choose problems to be exhibited to share with the whole class was a big challenge for Anna. She also found that children, when in the discussion of results on class work, would tend to attend to superficial features of the problems such as the handwriting (too big, mistakes in characters). Finally, children tended to make up problems that were similar and not creative in variety.

Telling stories. Anna was frustrated and came to meet her advisor who was also a teacher educator. They talked about alternatives and decided to make the lesson interesting before asking grade three children to pose their problems. Finally, Anna used the story to arouse interest in MPP. The strategy was, that the teacher started to tell a wonderful story but did not finish. The children asked what happened next. The teacher suggested children imagine what happened. When the story context was there, Anna used story completion to promote interest in making up problems together. Besides, Anna also found that she needed to give a hand to low-ability children just to help them start. By and by, she succeeded in letting children feel that they could control their learning and construct meaningful learning. Apart from learning mathematics MPP teaching also learned more about children's real life after school through the problems they made.

4.3.3 Knowledge of Content and Curriculum (KCC)

As given in 4.2 Anna referred to the math curriculum for grade three and studied students' outcomes then arranged them into a sequence of problems for use in teaching a certain topic (e.g. time, length, angle). For example, in *time*, she referred to the textbook of grade three and order the students' outcomes so that she could replace textbook problems on (1) Reading and telling what the date was, or telling how many Sundays in the month, (2) counting a certain number of days from a given date and telling the date, and, (3) calculate the total number of days between 2 given dates. The same applied to length, angle and other units in Table 1.

4.4 Reflections of an Action Researcher After Doing MPP Teaching

Anna found herself being able to act and to reflect on: the problem set up, being a facilitator, a time manager, and, being able to evaluate the effectiveness of her teaching.

Facilitator. The teacher provided a scaffolding for children to construct knowledge and was also responsible for leading whole class discussions. For example, prompting children to read written tasks and report what they found: triangles, quadrilaterals, vertices, angles, right angles…etc. Then children will not be distracted and concentrate on the mathematics content given in the tasks.

Time manager. Normally the teacher used 5 to 10 min to present a task. Generally, the presentation of MPP tasks for groups was easy to control but not so for individuals. When in groups they helped each other and the between-group differences were not high. However, for individual problem posing, the individual differences were so high, and time control was hard.

Evaluation of effectiveness. For example, when she considered children making similar problems, she did not think the MPP teaching was effective so she tried other means to make them pose creative problems. Therefore, the teacher acted as a person who constantly evaluated teaching, made self-reflections solved problems, and used their ruler to measure the effectiveness of their teaching. Through interviewing children, the teacher learned more about the activity she tried. The teacher asked if children liked MPP, (1) if they liked posed problems, and, (2) the reasons why.

"I like to write my problem because I can write it that way" (S1), "because it was fun" (S3), "can use our imagination" (S7), and "I can appreciate my friends because of the problems they posed during class" (S9). The reasons are, *"I like my posed problems because I like mathematics" (S6), "because they are good" (S4/S5). "I like my problem very much." (S2) and, "I do not know why", "I like my friends' problems compared to mine because they were harder" (S10)*

There was one child who preferred textbook problems because they were more difficult. When asked if he liked difficult problems, he nodded his head to indicate affirmative. The result was the same as the result in posing easy/hard problems given in van den Heuvel-Panhuizen et al. (1995).

5 Conclusion

What can Anna and Professor Leung learn from this action research? How could a teacher and teacher educator gain knowledge after teaching MPP?

Anna's voyage as a researcher made her encounter teaching problems and in resolving one after the other, she underwent *Groping, Transition,* and *Mature* stages. Most of the teaching problems were related to teaching math in general. However, there were few related to the use of problem posing in teaching. Initially, she did not know how to define "problem posing" to students in grade three who had no experience at all. Therefore, she asked students to "help the teacher make up problems, just like thinking problems for friends to take in an exam". Second, her professional growth over time included knowledge of math content and pedagogy content knowledge (Shulman, 1986), further divided into knowledge of content and students, knowledge of content and teaching, and, knowledge of content and curriculum (KCS, KCT, and KCC, Ball et al., 2008). She prepared class in advance and analyzed students' work after the lesson. This was habitual after she did action research. Third, Anna cultivated a habit of making reflections upon practice. When there was too little creativity in the posed problem by the students she would not say, "let it be." She figured it out and tried to resolve it as a teaching problem. Anna used magic bag as set-up and did storytelling to increase students' creativity in posing problems. She made a crown and said if the problem pleased the King, the King let the owner of the problem wear it for a minute.

However, the above problems did not appear only in teaching math problem posing. Anna used what she learned from walking through these three stages in her future teaching. For example, the teacher might waste time waiting for students to put up their work and be indecisive in selecting posed problems for discussion. She had a solution to this teaching problem during her action research. Anna invited students to write on a small whiteboard to present their work after the lesson. Then the teacher read afterward and selected work to discuss the next day. This would avoid including posed problems that led to discussion limited to the superficial feature of the problem like "*the handwriting was ugly*" or "*this does not look like a triangle*". For the other difficulties encountered, such as some quick students having nothing to do after finishing problem posing and interfering with other students Anna circulated drawing sheets for them to draw while waiting for friends to complete.

In Professor Leung's journal, the final brief notes were recommendations for another teacher who follow another MPP investigation:

1. Teaching math can be done via math problem posing (MPP) but preparation is important. The intended materials were built into the tasks. The teacher also needs to attend to classroom management before MPP introduction. The challenge is to know the math content and the students very well.
2. The teacher uses strategies to set up a task for MPP and makes it attractive. Manipulatives or story-telling can help to start. When children work on tasks, the teacher helps low-ability children and makes sure the high-ability children do not disturb those who are still working.

3. For discussion, do not attend to superficial features. Follow what other students talk about. Do not add unrelated content to the discussion. Noise is accepted when the whole class is discussed, as long as the noise is not obstructing students. Peer review is hard for children at this age and there is a need for the teacher to guide them or show them how
4. MPP educates not only on math but also on well-being. Students learn to share their thoughts and to respect others' ideas. Teachers may try MPP as homework such as posing problems and solving them at home. This homework can be used to improve family members' relationships (Leung & Wu, 2000).

Twenty years after her study she still kept in touch with her advisor; their mutual interests in problem posing and action research kept them together. Upon completion of Anna's study, she went to present results at a conference on action research and contributed a chapter for a book. Anna's chapter was given to Leung's graduate students to read and an online meeting enabled Anna to talk to graduate students to better their understanding of MPP. Conversely, when Anna was chair of academic affairs in the school, she invited Professor Leung to her school to do professional development sessions.

Part II ends here with actions by Anna (this chapter), which augment Wonder's development of MPP program (Chapter 9), Leung's teaching MPP directly (Chapter 7), and the use of MPP tasks and students' outcomes in bridging in-service teachers and prospective teachers (Chapter 8). The four cases explained how MPP enhanced professional development.

References

Ball, D. L., Thames, M. H., & Phelps, G. (2008). Content knowledge for teaching: What makes it special? *Journal of Teacher Education, 59*, 389–407.

Borba, M. C. (1994). High school students' mathematical problem posing: An exploratory study in the classroom. Paper presented at the Annual Meeting of the American Educational Research Association. New Orleans, LA.

Butts, T. (1980). *Posing problems properly*. In S. Krulik & R. E. Reys (Eds.), *Problem solving in school mathematics* (pp. 23–33). National Council of Teachers of Mathematics.

Clarke, D. (2008). The mathematics teacher as curriculum maker: Developing knowledge for enacting curriculum. In In P. Sullivan and T. Wood (Ed.). *Knowledge and beliefs in mathematics teaching and teaching development* (pp. 133–151), Sense Publishers.

Hill, H. C., Ball, D. L., & Schilling, S. G. (2008). Unpacking pedagogical content knowledge: Conceptualizing and measuring teachers' topic-specific knowledge of students. *Journal for Research in Mathematics Education, 39*(4), 372–400.

Leung, S. K. S. (2013). Teachers implementing mathematical problem posing in the classroom: Challenges and strategies. *Educational Studies in Mathematics, 83*(1), 103–116.

Leung, S. S., & Wu, R. X. (2000). Sharing problem posing and problem solving at home through diary writing. *Australian Primary Mathematics Classroom, 5*(1), 28–32.

Lincoln, Y. S., & Guba, E. G. (1985). *Naturalistic inquiry*. Sage.

Ministry of Education of Taiwan. (1993). *Curriculum standards for national elementary schools in Taiwan* (pp. 91–132). (In Chinese)

Ministry of Education of Taiwan. (2008). *General guidelines of grades 1–9 curriculum for elementary and Junior High School Education* (in Chinese).

Ministry of Education of Taiwan. (2018). *Curriculum guidelines for the 12-year Basic Education Elementary School, Junior High School, and Upper Secondary School-The Domain of Mathematics* (in Chinese).

National Council of Teachers of Mathematics. (1991). *Professional standards for teaching mathematics*. Author.

Shulman, L. (1986). Those who understand: knowledge growth in teaching. *Educational Researcher, 15*(2), 4–14.

Smith, J. P., & Barret, J. E. (2017). Learning and teaching measurement: coordinating quantity and number. In J. Cai (2017). *Compendium for research in mathematics education* (pp. 355–385). NCTM.

Smith, M. S., Silver, E. A., & Stein, M. K. (2005). *Improving instruction in geometry and measurement. Using cases to transform mathematics teaching and learning*, vol. 3. Teachers College Press.

Tsubota, E. (1987). *On children's problem posing (grade 1 to 3)* (In Japanese).

Van den Heuvel-Panhuizen, M., Middleton, J. A., & Streefland, L. (1995). Student-generated problems: Easy and difficult problems on percentage. *For the Learning of Mathematics, 15*(3), 21–27.

Winograd, K. (1990). *Writing, solving and sharing original math story problem: Case studies of fifth grade children's cognitive.*

Supporting Teachers to Promote Students' Mathematical Problem Posing: Commentary on Part II (Teachers' Professional Growth)

Edward A. Silver

I was grateful when Shuk-kwan (Susan) Leung invited me to write a commentary for a section of her book, *Mathematics Problem Posing in Action: Students' learning, teachers' professional development and parental involvement*, which provides highlights of her multi-decade program of research and development aimed at understanding and promoting mathematics learning in Taiwan with and through problem posing. The invitation to contribute to her book was especially delightful because I witnessed the dawning of Susan's interest in mathematical problem posing (MPP) when I served as her academic advisor and dissertation supervisor at the University of Pittsburgh more than 30 years ago.

Leung's Approach to Mathematical Problem Posing

Her dissertation (Leung, 1993) reported what was at that time one of only a handful of research investigations into mathematical problem posing. In her study, she examined the mathematics problems posed by a group of prospective elementary school teachers. She analysed the cognitive and mathematical complexity of their posed problems, and she probed the relationship between the problems they posed and their performance on tests of creativity and mathematical knowledge. She found essentially no relationship between their problem posing and their scores on a test of creativity, but she did report a strong relationship between the mathematical knowledge of the prospective teachers and the complexity of the problems they posed.

It was clear from her initial study that Susan recognized the central role of teachers in mathematical problem posing. Yet, as Jeremy Kilpatrick (1987) noted at the beginning of his seminal chapter on problem formulation, students rarely have

E. A. Silver
University of Michigan, Ann Arbor, MI, USA
e-mail: easilver@umich.edu

school experiences that cultivate an orientation toward posing mathematics problems (Kilpatrick, 1987). This lack of problem-posing tasks in curriculum materials had at least two unfortunate consequences for teachers. One is that they are very likely to lack personal experience with mathematical problem posing in their own school education. Another is that the curriculum resources available to them to support their instruction are unlikely to provide useful, consistent support for them to implement problem posing in their classrooms. Susan recognized immediately that changing this normative feature of school mathematics required that teachers increase their familiarity and facility with mathematical problem. Fortunately, her dissertation study findings suggested that mathematical problem posing was something teachers could do and need not require exceptional creativity.

Upon completing her doctoral degree, Susan began a very productive career as a mathematics teacher educator (MTE) in Taiwan, where she has taught students and worked with teachers and parents over 30 years. Across the years, she vigorously pursued her interest in and commitment to mathematical problem posing, publishing many papers based on her research and practice initiatives. The most recent mathematics curriculum revision in Taiwan includes mathematical problem posing as an important feature of students' school experience and urges teachers to incorporate MPP into their instruction. Though overt support of MPP by the Taiwan Ministry of Education is a major step forward, Shuk-kwan (Susan) Leung has understood from the beginning that it is teachers who are the ones who must make MPP happen in school and that teachers would need considerable support in making such a change to their practice. Thus, she has worked with steady determination and creativity to promote MPP in Taiwan. Her efforts have culminated with the publication of Mathematics *Problem Posing in Action,* which summarizes key aspects of her work with students, teachers and parents specifically related to her quest to infuse MPP into classroom instruction in Taiwan.

Leung's Approach to MPP Teacher Professional Education

My assignment was to prepare a commentary on Part II of this book, in which we find chapters that focus on Leung's professional development work with teachers. This section contains four chapters, each chronicling a different professional development approach, different teachers, and different settings, and all focused on mathematical problem posing.

The chapters in Part II provide evidence that Susan approached her MTE work with considerable care, determination and creativity. Across the four chapters we see the impressive breadth and depth of Leung's MTE work as she steadfastly supported teachers to undertake MPP in their classrooms.

We find here accounts of her work with prospective teachers and current teachers—at the elementary, lower secondary and upper secondary levels. Moreover, these chapters illustrate her use of a variety of methods to support teacher professional learning. Some approaches are likely to be quite familiar to readers, such as teaching prospective teachers in the same manner as if they were schoolchildren. Other approaches are likely to be less familiar, such as teaching classroom groups of schoolchildren herself to demonstrate her ideas in action for prospective or in-service teachers to observe and discuss with her. She acknowledges and illustrates that she approached her work from a variety of perspectives, including as a university professor, as an action researcher, as a colleague and co-teacher, as a friendly critic of teachers, and as a teacher willing to have her own practice inspected and critiqued. And throughout we see her careful attention to MPP as an instructional approach, not as a faddish gimmick but rather as a clever way to engage students and enhance learning of the mathematics at stake in a lesson.

In Chapter 7, we see Leung's use of the so-called fishbowl method to promote professional learning. In this approach, which is adapted from the widely used surgical theatre used in traditional medical education. a skilled practitioner (teacher or surgeon) "performs" in an authentic setting (classroom or hospital operating room) under the watchful gaze of learners (teachers or medical students). To create this type of learning opportunity, Susan "borrowed" an intact class of students and conducted a mathematics lesson on the topic of equation solving. She conducted three different lessons—each at a different grade level with a different group of children, a different group of teacher-observers, and a grade-level appropriate use of equations. What remained constant across these settings was the "fish" and her use of MPP as a key aspect of her instructional lesson.

In Chapter 8 the context shifts to curriculum development and instructional design, as Leung describes her direction of a university-based team of investigators in the development of a set of more than 50 MPP tasks intended for use in elementary school classrooms. More than 100 in-service teachers tried versions of the tasks in their classrooms and collected data that was analyzed by the development team. This cyclic process of task development and trial usage, leading to data collection and analysis, and including teacher feedback and reflection was repeated several times and served not only as a way to improve the task development but also as professional development for the teachers who were not yet familiar and comfortable with using MPP in their instruction. We see several of the rich mathematical tasks the team designed for use by teachers and see examples of the interesting and substantive ways students responded to these novel tasks. Also included in this chapter is a description of how the tasks and students' work samples were used to enrich Leung's work with prospective elementary teachers.

In Chapters 7 and 8, we saw examples of Susan's work with relatively large groups of teachers. In Chapters 9 and 10 the focus shifts to cases of her intense engagement with individual teachers. These chapters provide "existence proofs" of the feasibility of elementary teachers enacting MPP as a regular feature of their mathematics instruction, with beneficial consequences for students.

Chapter 9 reports an interesting case of a teacher (Ms. Wonder) who uses MPP as a key feature of her instruction in elementary school over 12 weeks. Leung regularly observed Wonder's classroom lessons and met with her weekly to discuss her reflections and impressions of progress. Together they evaluated the outcomes of this instructional experiment and drew lessons to guide future instruction and research. In this instance, Leung's MTE work was embedded in her participation as a participant-observer and co-investigator with a teacher engaging in an ambitious instructional reform initiatives.

Part II concludes with this chapter, in which we meet another teacher (Anna) undertaking action research in her third-grade classroom. Anna had five years of teaching experience, and she was pursuing a master's degree as Leung's advisee. For her thesis, Anna pursued an action research project on MPP for a full semester. The teacher/researcher evaluated the activities and outcomes using a variety of data sources, including lesson plans, classroom videotapes and observation notes, interviews, and reflection notes. The chapter details some remarkable examples that illustrate what Anna learned in and through her action research. For example, we see evidence of changes in her beliefs and in her approach to formulating and solving problems of her teaching practice that she viewed as applicable not only to this MPP episode but also to a broader set of issues in her future teaching. Leung also shares some lovely examples of what she learned as Anna's companion and co-investigator on this action-research adventure and how she applied Anna's lessons to her future work with teachers.

Concluding Remarks

Mathematical problem posing was the topic of a plenary lecture I presented about 30 years ago at the 17th annual conference of the International Group for the Psychology of Mathematics Education (PME). In that talk, subsequently published as a paper in *For the Learning of Mathematics* (Silver, 1994), I suggested some reasons why mathematicians and mathematics educators should be interested in problem posing, and I summarized a few research studies (including Susan's dissertation study) related to students' mathematical problem posing. Though the research base at that time was rather thin, I hoped that a PME plenary lecture that focused on mathematical problem posing would elevate its visibility among mathematics education scholars. Since then, I am delighted to see that professional interest in mathematical problem posing has grown considerably.

Although problem posing is a relatively new topic in mathematics education, it is one that appears to have captured quite a bit of attention in recent decades. A recent search of Google Scholar for the topic "mathematics problem posing" yielded about 400,000 citations. Although that number is only about 10 percent of the total citations for "mathematics problem solving," a check of citations since 2000 yielded almost identical amounts for the two topics (about 18,000). Thus, mathematics problem posing appears to be capturing an equal amount of attention lately when compared to the more established topic of mathematics problem solving.

From the dramatic growth in attention to mathematical problem posing it is evident that mathematics educators agree on the importance of mathematical problem posing. Unfortunately, the research done to date has generated an uneven understanding of aspects of mathematical problem posing, with far more known about the nature of problem posing and problem-posing processes and far less known about how to incorporate it into classroom instruction.

Shuk-kwan (Susan) Leung's book, *Mathematics Problem Posing in Action: Students' learning, teachers' professional development and parental involvement*, is a welcome and much needed addition to the existing corpus of scholarship on mathematical problem posing. In Part II of this book, she offers detailed accounts of MPP as enacted by teachers and provides insights into the kinds of supports that might be required for teachers to undertake and successfully enact this innovative form of mathematics instruction. The book serves not only as a valuable resource for mathematics teachers, mathematics teacher educators, and mathematics education researchers, but also as a fitting capstone to her 30-year career as an advocate for mathematical problem posing.

References

Kilpatrick, J. (1987). Problem formulating: Where do good problems come from? In A. H. Schoenfeld (Ed.), *Cognitive science and mathematics education* (pp. 123–147). Hillsdale, NJ: Lawrence Erlbaum.

Leung, S.S. [1993] *The relation of mathematical knowledge and creative thinking to the mathematical problem posing of prospective elementary school teachers on tasks differing in numerical information content.* Unpublished doctoral dissertation, University of Pittsburgh.

Silver, E. A. (1994). On mathematical problem posing. *For the Learning of Mathematics, 14*(1), 19–28.

Part III
Parental Involvement

Chapter 11
Mathematics Problem Posing Homework for Parents and Children: Two Case Studies on Number and Mensuration

1 Introduction

Parental involvement (Section III) is the third branch of the tree planted by the same math teacher educator (MTE). Here, she named herself Mytle. Here, she worked with parents, which was not a new experience.

Mytle went to teach her sons (when they were in elementary school) as a parent (not MTE) and documented her teaching stories in a book for parents and teachers (Leung, 2008). It included her motivation and preparation of simple teaching aids, the intention to reinforce math lessons taught by the teacher, and, a story of what happened. Voices of children and Mytle were carefully included, to convey to teachers or parents how an adult can interact with children through a math activity designed for a certain age group. A grade one teacher, Wu, from Mytle's team, led a group of parents to read and met on Friday afternoons to talk about how children learned. Leung (2008) was used in a parents study group given by a grade one teacher who made parents learn math in the same way their children learn math. After one term, the study group enhanced parents' learning in math (Leung et al., 2012). There were positive effects on children's math learning, attitudes, interaction with parents, and, also parents-teacher communication. The findings also indicated parents need advice in supervising math homework.

Over the past 30 years of Mytle's career, Taiwan has had different waves of curriculum change. In the latest curriculum standards (MOE, 2014) the official document emphasized the part on parental involvement. In MOE (2014, p. 22), it says: (1) Curriculum implementation requires support and participation from parents. Schools should encourage parent associations to establish parent-learning communities or parent-teacher co-learning communities to improve educational knowledge among parents, strengthen collaboration between parents and teachers, and support students' effective learning and adaptive development. (2) Schools should regularly invite parents to attend various courses or teaching activities or ask them to participate in teaching lessons that welcome the participation of other faculty members. This

S. S. Leung, *Mathematics Problem Posing in Action*,
https://doi.org/10.1007/978-981-96-8480-9_11

encouraged parents to care more about their children's classes and the implementation of curricula and teaching practices. Accordingly, a positive communication channel between parents and the school culture can be established to promote co-learning among teachers, parents, and students.

At the time of the studies on parental involvement in this section, Mytle continued her prior work and responded to emphasis on the curriculum standards. Her preliminary attempt was on math problem-posing homework. She started two case studies. For case one, the attention was given to: (1) types of problems posed and solved by dyads together, and, (2) types of interactions and how dyads considered real-life, interests and appreciation. For case two, the class teacher first taught math problems posing in class then followed up with parents posing problems at home. The research questions were: (1) types of problems posed during in-class MPP instructions and attitudes of students, (2) interactions of parent and child when doing MPP at home and their views about completing MPP homework together at home.

2 Background Literature for Actions

2.1 Parental Involvement in Children's Math Learning

Parents are children's earliest teachers. Involving parents in children's growth and well-being catches the attention of research and practice nationwide. Projects on parents ranged from caring for infants and transition from home to pre-school. However, as children attend elementary school, they continue to need parental attention. In international studies on assessments, such as PISA, results confirmed the importance of parental involvement. (OECD, 2020). Parental involvement can be in the form of checking homework, or, completing written work together with children. In Taiwan, even for families where both parents take up full-time jobs, parents spend time on their children's homework regularly and communicate with teachers using a teacher-parent communication book.

Although math problem posing is gaining its importance in research and practice and teachers teaching math in school might also introduce MPP, how parents attended to children using problem posing was not much researched. Leung and Wu (2000) discussed possible actions a teacher can take and extended MPP to parents at home. In this study, a simple MPP written task was used to invite parents to pose problems with children at home. If children could not write the Chinese characters, then they wrote phonetic symbols of the characters to express in writing. The teacher named "worksheet" a "math diary", and with this name, she succeeded in getting promising work shared from the home. Besides, "homework" was not used nor mentioned. She found that MPP could focus on both math learning and parent–child interaction. The types of returned written work were in multiple forms: Posing and solving together, child posed and solved, adult posed then child solved, and finally, child posed then adult solved.

In a study by Williams and Williams (2021), nine English primary schools used carefully designed homework to generate specific forms of parental involvement in children's learning and found mathematics homework enjoyable under two conditions. First, when homework was situated in everyday applications of mathematics. Second, if teachers convinced parents that they can help regardless of their own perceived degree of mathematical competence. In addition, children who learned with parents were then more likely to recognize the usefulness of mathematics as a life skill and the value of feedback. In addition to real life, interesting homework may foster children's mathematics talk, and increase self-efficacy and completion rates. Finally, the homework included the part on appreciation: having adults and children enjoy the presence of each other in completion.

In the latest version of Taiwan curriculum standards (MOE, 2018), the part on problem posing is also emphasized. However, there is no specific way to integrate this activity in school or extend it to home. What can a class teacher do to encourage parents to do so? The teacher can design worksheets for children to bring home and try to complete MPP in written tasks (Leung & Wu, 2000; Williams & Williams, 2021). In problem posing, families might add feelings and activities of activities after school. Adding math problems posed as homework, might allow investigators to address real-life, interests and appreciation mentioned in Williams and Williams (2021).

2.2 *Mensuration and Number Learning*

Mensuration and number were listed under the first strand (among four) in the Taiwan math curriculum guidebook (NAER, 2020), and occupy most of the instructional time in learning math for the elementary school.

Mensuration. In the earlier part of schooling, students all over the world are introduced to measurement via the quantities of everyday experience: length, capacity, time, weight, temperature, volume, and soon area and angle (van den Heuvel-Panhuizen & Buys, 2008). In a home environment, parents and children come across a lot of problems with measures. Smith and Barrett (2017) discussed the learning and teaching of measurement and included that part on coordinating quantity and number. For the part on learning mensuration in Taiwan, the order is: Money, time, length, weight, area, volume, capacity, angle, and, speed (NAER, 2020). The order followed how familiar children experienced the quantities in real-life situations. The list included "money" as a measure to be learned in school. However, the measure of money was not represented in proportion by weights or lengths. Say, a ten-dollar coin is not ten times the weight of a one-dollar coin. Young children experience using money in real life or learn in school using teaching aids like fake money. According to the curriculum standards document (MOE, 2018), and, in grade one, students learn about coins ($1, $5, $10, $50). In grade two, they come across paper money ($100, $200, $500, and $1000). When in grade three they learn to make purchases that might involve a money conversion. For example, they learn how to handle units at two levels (e.g., tens and ones, hundreds and tens).

Number. For the part on number learning in Taiwan, the strand was subdivided into Numbers and Operations (see MOE, 2018). Under number, there are whole numbers, fractions, decimals, and rounding numbers (after approximation). Under operations, there are addition, subtraction, multiplication, and division. By grade four, a student already learned all four algorithms on whole numbers; including one-step problems on each operation, and, also, two-step problems on mixed operations. Also involved calculations, completing a fill-in-blank arithmetic expression, or finding solutions to word problems.

It is worth investigating, when teachers invite parents and children to pose problems using worksheets concerning money or numbers, what kinds of problems will be resulted.

3 Method

3.1 Participants

The investigator is a math teacher educator. Mytle was a parent and as well a resource person who taught in her son's school. Lessons on the teaching were given in two books (Leung, 2008, 2015). She taught in a university as a math teacher educator; and, organized of family math camps in an elementary school near the university. In this chapter, she included participants from families for doing written tasks on MPP.

Participants were parents and children from grade 1 to grade 4. *Case One.* Participants were 54 parents and 54 children from grade 1 to grade 3. At the time when the study was conducted, the 108 syllabus (MOE, 2018) was implemented from first grade to third grade. One class from each grade was involved in the data collection procedure. *Case two.* Participants were parents and children from a grade four class of 27 children with 13 boys and 14 girls.

3.2 Design of Materials

Materials were in alignment with math lessons but given in the form of homework. *Case One.* The investigators named the data collection booklet, "After-dinner-math" (also the name of a book by Leung, 2015). The "After-dinner-math" booklet included problem posing and feelings about family math time after dinner. For parent–child problem posing, there were two parts: "Problem posing and solving together" and "You-pose-I-solve". The investigator studied textbooks, curriculum guides, and types of problem posing tasks (Leung, 2013). They developed MPP tasks as in Table 1.

Table 1 Overall design of materials by strands and by MPP types

Grade	Name of task	Name of units	Strand		Types	
Grade1	1. Storeroom	Understanding shape		S		F
	2. Eggs	Add/subtract	N			F
	3. Pocket Money	**$1, $5, $10 (Money, coins)**	**N**		**T**	
	4. Snack	Take away /add/subtract	N			F
Grade 2	1. Chinese New Year	Plane figures		S		
	2. Fake money	**$200, $500, $1000 (Paper Money)**	**N**		**T**	
	3. Calendar	Two-digit subtract	N			F
	4. Candy & basket	Year, month, day	N			F
Grade 3	1. Add/subtract	Count to 10,000	N		T	
	2. Angle measure	Angle, squares		S		F
	3. Purse	Multiplication	N		T	
	4. Mask, alcohol	**Add/subtract up to 4-digit number (Money conversion)**	**N**			**F**

Strand N = Number and mensuration, Strand S = Shape and space,
MPP types T = Text, MPP types F = Figure or Table

As in Table 1, there were four tasks in each grade in Number and mensuration (N) or Shape and space (S). The MPP types were either a task with a given text (T) or with a figure or table (F). One task from each grade was marked in **bold** to represent the reported target topic here. It was about money (N) and the three tasks were about coins, paper money, and, money conversion.

In this "After-dinner-math" booklet, there was a catchy front page of instructions to child (left) and parent (right) (Fig. 1). The four tasks for posing and solving followed. The booklet consisted of an end page for responses on real life, interests, and appreciation (Fig. 2).

Case two. The MPP tasks were in paper and pencil format. The MPP types were classified according to "*given*" (see Leung, 2013). They were named according to the six types of given: Given an expression, a text, a figure, a table, an answer, and, a problem. The three math topics were chapters taken from the textbooks.

For *Fill-in-blanks arithmetic involving multiplication and division*, students needed to know each of the two operations was inverse of the other. Then, they tried to find the number that should be put into the blanks (i.e. the missing number) so as to make the expression correct. If the expression is $5 \times 4 = 20$, one of the values will be taken away and replaced by a blank or bracket. Students need to read carefully and find the number using facts from multiplication and division.

For *vertical multiplication and division*, the multiplication is a three-digit number to a two-digit one and given in a word problem. As for division, the problem will

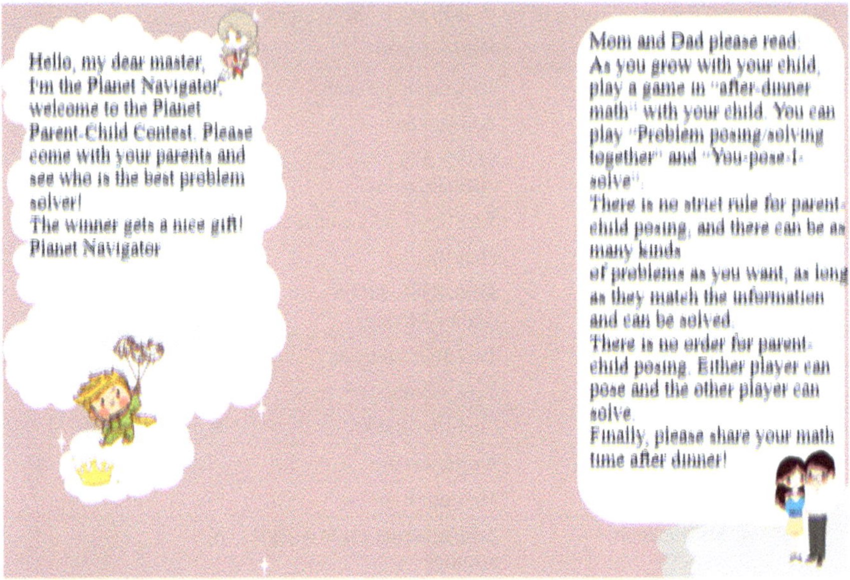

Fig. 1 Front page of "After-dinner-math". (Grade 1 to Grade 3)

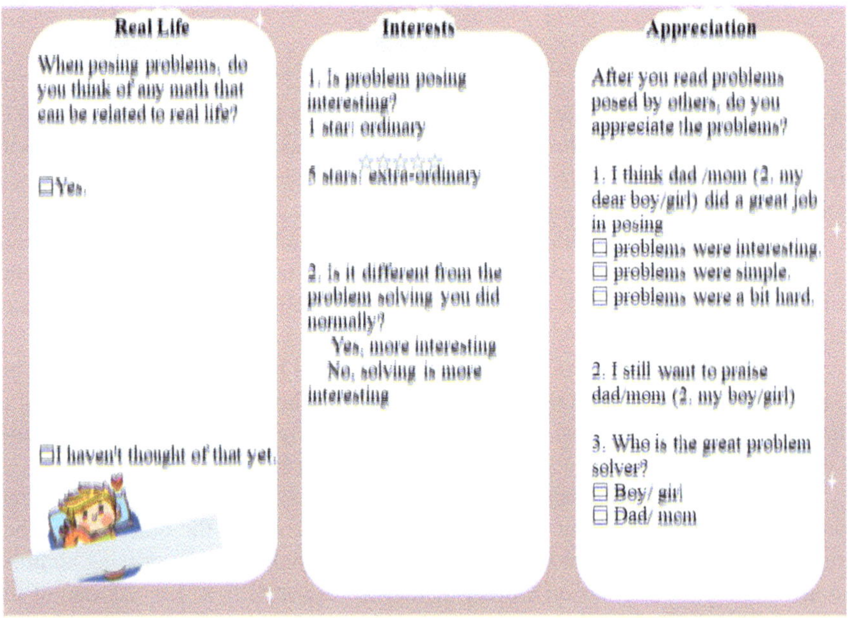

Fig. 2 End page of "After-dinner-math". (Grade 1 to Grade 3)

"Let us share today's math"

Dear student(s): Refer to the chapter on *<u>Fill-in-blanks arithmetic involving multiplication and division</u>*, *please share with your parents(s) what you already learned today in school. After they know about it, make up similar or different problems with your parent(s) and put the problem below. Please share the results here.*

The problem we make together is as below.

(My solution is here)

We make a very different problem for friends to solve (not solved yet).

Dear Parent(s): This worksheet is to help your child revise the lesson and practice solving word problems. In this way, you get to know what they learned in school. Please spend some time and pose problems with your child. Please share what happened.

Signature of parent (Date:)

Fig. 3 "Let us share today's math". (Grade 4)

be either a three-digit dividing a two-digit or a four-digit number dividing a three-digit number. The operation would lead to an exact division or a division with the remainder.

For *computation involving four algorithms and its rules*, repeated subtraction first then using rules from left to right, parenthesis first, and, also the rule governing working on multiplication and division over addition and subtraction.

With the above MPP types (n = 6) and math topic (n = 3), Mytle and the teacher developed as materials for grade four children and parents. Note that they did not mention "homework". The worksheet started with *"Let us share today's math"*, and, posing problem was changed to "making up problems". (Fig. 3).

3.3 Data Collection and Analysis

Case one. The investigator collected work from one class from each grade (1, 2, 3) forming a convenient sample (n = 54). The tasks were on the money: coins (grade

1), paper money (grade 2), and conversion of units in money (grade 3) respectively. Parents and children were to pose problems according to the information. Multiple forms followed that of Leung and Wu (2000) were used. One task was when a parent posed problem(s) for a child to solve, then, the child posed for a parent to solve (Grade 1). There was another task for posing and solving together (Grade 2). Finally, one child posed problem(s) for a parent to solve, and then a parent posed for a child to solve (Grade 3). *Case two.* Problem-posing instructions were given in parent–child problem-posing worksheets. Data collection included MPP worksheets (in-class, at home), mathematics attitude measures, students' interviews, parents' interviews, teachers' journals, and videotapes of math lessons.

Data was analyzed using mixed methods (Creswell, 2009), including descriptive statistics and qualitative methods. *Case one.* The investigator referred to worksheets and did coding on types by using Leung and Silver (1997) and coded as "1" to "5": non-math (1), math (2), implausible (3), insufficient (4), and Sufficient/Extraneous (5). A reliability check was done on 20% data at random. Later, all problems from Category 5 were further classified as number and mensuration (N), space (S), relation (R), and, data handling (D). For the data on sharing feelings on math time after dinner, the analysis was done qualitatively. First, in real life, responses were divided into math-related and non-math-related; then those common to all three grades and those given by each grade. Second, on interests, responses were divided into feelings, recall MPP process, feedback to the investigator, and vote of thanks to each other. Third, on appreciation, what parts to appreciate from each other, and, which of them was the best problem solver and the reason. *Case two.* Similar to case one data analyses included firstly, data coding of the posed problem by Leung and Silver (1997). A reliability check of coding from worksheets was done by a colleague who also taught grade 4. Data on mathematics attitude measures were analyzed quantitatively. Triangulation on data from students' interviews, parents' interviews, teachers' journals, and videotapes of math lessons. An instrument developed locally was used to measure math attitudes in six scales: Confidence, motivation, habits, habits, self-conception, communication, and interaction (Hsu, 2001). Qualitative analyses were done on data on parent and child interactions.

4 Results and Discussion

4.1 Case One (Grade 1 to Grade 3)

Results were in two sections, according to the research questions (types of problems, interaction). There were three forms of having parents and children pose and solve (Leung & Wu, 2000) and for three tasks concerning money in each grade (see Table 1 again).

4.1.1 Types of Posed Problems from Parent–Child Pairs

The investigators reported types according to sufficiency (Leung & Silver, 1997) and math content structure (Heller & Greeno, 1978).

Sufficiency. The first part of "*Mathematics After Dinner*" referred to the types of problems given by parent–child. The average number of problems posed per family was 5.7, 5.2 to 10.4 (grade three having the most). Classification of sufficiency was done according to Leung and Silver (1997). Together with parents, children posed 96.7% *sufficient* problems (351 out of 363). For *insufficient* problems, two were missing goals (grade one) and 10 were missing givens (five from grade 2, another five from grade 3).

Math content structure. All the above *sufficient* problems were further categorized according to math content structure.

Parent-Pose-Child-Solve then Child-Pose-Parent-Solve

Example from Grade One, Marked Price ($n = 24$).

Nezuko has $15, her brother Tanjirou has $17, an ice cream costs $8, a chocolate costs $12, four pens cost $6.

The task was designed to go with the unit on $1, $5, $10 (Money, coins), and also a unit taught before called "take away and addition and subtraction".

The posed problems were on Numbers. There is substantial literature on problem solving, on how children solve addition and subtraction word problems (Heller & Greeno, 1978) but less on how they pose them (English, 1998). Children's understanding of situations matching to operate symbols such as " + " or "−" was limited. In solving word problems, the difference in categories of semantic structures (CHANGE, COMBINE, and, COMPARE) might made problem solving different. Children's percentage of correct in solving different categories suggested the varying amount of difficulty levels in understanding the word problems (Nesher et al., 1982). For example, CHANGE was found to be easier than COMBINE, and, COMBINE was not as difficult as COMPARE.

How about problem posing?

A total of 29 parents' posed problems was resulted. There were 16 subtraction CHANGE problems (find the amount left) and the remaining 5 involved mixed operations (*How much should they pay for 2 chocolates and 4 pens?*) Besides, there were 8 addition problems with COMBINE as semantic structure, and unknown sum like (*How much altogether?*). However, there was none on COMPARE. Therefore, the results on problem posing were similar to results given in prior research on problem solving (e.g. Nesher et al., 1982). CHANGE was found to be easier than COMBINE. And none in COMPARE meant not common. They were most difficult to pose; parents and children were not familiar with the COMPARE category of semantic structure.

Among the 24 children's posed problems, 15 were subtraction CHANGE problems (find the amount left). There were 8 addition problems with COMBINE as semantic structure, including 6 on the unknown sum (*How much altogether?*). Finally, the remaining one problem appeared to be a division problem, "*One ice cream costs $8, how many ice cream can he buy using $24?*" In problem posing, it was common to result in a posed problem not matching the difficulty level of the child who posed it. In this example, the parent accompanied this child (in grade one). They did backward counting and the answer was, "Can buy 3 ice-creams."

There were many other kinds of problems about price, whether posed by parents or children. Except for one multiplication problem, the others were having semantic problems that appeared in most textbook. There were even word problems that required two steps or division problems.

Dyads Posed then Solved

Example from Grade Two (n = 11)

A figure with paper money: *fourteen $200 (eight of them were crossed *out); five $500 (two were crossed *out); and three $100 (1 crossed *out).*

The task was designed to go with the unit on $200, $500, and $1000 (Paper Money), also a unit taught before called "add on or take away and addition and subtraction". There were different bank notes, 200's, 500's and 1000's and some of them were crossed out by strokes. It was left for children and parents to figure out what the strokes across them meant. This problem was also designed for the units on two-digit subtraction.

Among the 11 dyads who posed problems they solved together, there was one problem on measure, about paper money: Find the number of paper money that were crossed out. how many $200? How many $500? How many $1000? The remaining 14 problems were on computations. There was a total of 6 having semantics structure on COBMINE (add), with unknown results (the sum). "*Find the total amount he spent?*" The remaining 8 had semantics structure on CHANGE (subtract). Most of them (7) also asked for the unknown result. "*Dad has $8300 and paid a tuition fee of $4000. Find the amount of money left?*" There was one example with an unknown change: "Mom has $6900 and found $3900 after shopping. How much did she spend?" This problem was the odd one out.

In problem solving, "unknown change" was found to be more difficult than "unknown result" among different types of CHANGE problems (Heller & Greeno, 1978). Here, the result in problem posing was the same as in problem solving, if we take the fact that more frequently means easier. There was only one "unknown change" problem posed but a total of 7 "unknown results" problems were posed.

The Child Posed for the Parent to Solve (then Reverse)

Example from Grade 3 (n = 19)

"Mom went to the supermarket to buy masks and alcohol. Three packs of masks cost $36, one bottle of alcohol costs $72. She spent $288 in total."

The task was designed to go with the unit on the conversion of units (Money), also a unit taught before was "addition and subtraction".

The text gave a supply of information on the unit price for masks and alcohol and the total expenditure. This item was designed to go with the unit on "Introduction to Multiplication" and children already learned to add/subtract. All the posed problems belonged to the Number and computation category ($n = 34$). Parents posed 16 problems while children posed 18 problems. Parents' posed problems were on single operations like addition ($n = 4$), subtraction (2), and division (3). The remaining problems were two-step problems (7); involving a combination of two operations from the four algorithms. Children's posed problems included reading information (1), comparing two numbers (3), single operations like addition (3), and, subtraction (6). A few were two-step problems (3); involving a combination of addition and subtraction.

Example of a Single step: Addition. *Find the cost of buying three packs of masks and one can of alcohol.* Multiplication. *One can of alcohol cost $72. How much for 10 cans?* Division. *Three packs of masks cost $36, how many packs can a mom buy using $288?* Examples of Multi-step: Mixed operations. *If 9 packs of masks are packed in one box, and three cans of alcohol are packed in one box and Dad wants to buy two boxes of masks and one box of alcohol, how much will it cost?*

4.1.2 Parent–Child Interaction

The second part of the "After-dinner-math" worksheet attended to parent–child interaction (life, interest, appreciation). After the parents and children completed problem posing, they made efforts to think about the life application related to mathematics in the process, the fun of MPP between parent and child, and write down their appreciation of each other's works, and selected a great problem solver among the two players. This section included results on *real life*, *interest*, and *appreciation* as in a study on parents' problem solving homework (Williams & Williams, 2021).

Real life. "*When posing problems do you think of any math related to real life?*" They wrote "yes" and shared examples on *money, food, time, and games.* (1) *Money* was a kind of mathematical application, such as shopping, household income and expenditure, and lucky money received during Chinese New Year. (2) *Food* was considered to make problems real-life. Distributing food equally to each person was a division application, and, portions (part of a whole) left after the food was cut was a fraction application. (3) *Time* was another. Time described the changes of things, and it was a concept of quantity and actual measurement. For example, parents posed problems on dates using a calendar. (4) *Game* is in real-life of children, and, it was often necessary to use mathematics. In board games like Monopoly, a child counts the amount of money. In playing ball games, they record scores to determine the winner.

Degrees of Interest. For the first question, parents and children were asked to rate the degree of interests of the posed problem (full score is five stars). The scores were: first grade: 4.4 stars, second grade: 4.5 stars, and third grade: 3.4 stars. In terms of children's preference for problem-solving and problem posing, most of them preferred problem posing. Only two first-grade children, one second-grade child, and one third-grade child preferred solving; and, one second-grade child liked both. Parents and children expressed their moods and feelings when posing problems. Some children feel that "*it is very interesting to give the topic to their parents*", and "*I will think about it for a long time before I come up with a problem.*" In doing After-Dinner-Math together, they showed gratitude too. "*Thank you, Mom, for helping me with my homework.*" Parents showed interest in the process of posing and analyzing the work of children. For example, one said, "*Through posing and solving, you can understand the child's weaknesses and then to strengthen the deficiencies.*" Another parent said, "*We can use our brains together during the holidays. How nice?*".

Appreciation. First, parents and children are asked to rate each other's work. The options are *"the question is very interesting", "the question is simple", and "the question is a bit hard"*. There were 27 who thought their parents' work was *very interesting*; 14 *simple*; and, 17 *a bit hard*. On the other hand, all but one parent thought that children's works were *both very interesting and simple*. Only one first-grade parent thought that the children's work was *a bit hard*.

Second, some children appreciated the creativity of their parents' works and thanked their parents for accompanying them to formulate the questions and to find answers, like "*I think mom knew the answer already, but she pretended not and worked very hard with me.*" In return, some parents praised and encouraged their children; added that the children integrated with real-life materials and were flexible in writing; and, gave their children feedback. Parents admired the effort of children after school. "*Betty was struggling, although the activity cut down her time for TV and snacks.*"

Third, parents and children were asked to choose a great problem solver and wrote the reason to tell the teacher and friends. A total of 6 groups considered both were great problem solvers. "*No one is better than the other*", and "*We cannot decide as we were equally great*". There were 23 parent–child groups who thought that children were great problem solvers. The reasons were affective such as being confident ("*He told me he could and urged me not to help*"), and having endurance ("*I saw him erasing several times, still trying*"). Also, cognitively liked having strategies (*"He remembered to use the numbers to check, and he did twice."*) and knew mental arithmetic ("*She already had the answer without using paper and pencil.*"). Finally, 25 groups felt that the parents were great problem solvers. The reasons were affective ("*She was smart in prediction and we really completed it in 10 min*"), or cognitive ("*She was fast*", "*accurate*", "*showed me two methods to find the answers*").

4.2 *Case Two (Grade 4, N = 27)*

4.2.1 Results of MPP Instructions on Children's Problem Posing

Responses from MPP

From eleven lessons in three chapters: fill-in-blanks arithmetic (3 lessons), multiplication and division (4 lessons), and four basic algorithm (4 lessons), a total of 297 problems were posed (27 children, 11 lessons) (Table 2).

As above, children's posed problems added to a total of 297 problems.

First, there were six Not-a-problem ($n = 6$, Category 1)

e.g. A dog cost $2000. There were five.

They were posed by students from the lower 27% of the class. After the third lesson on problem posing, they could pose sufficient problems.
Second, there was zero non-math problem ($n = 0$, Category 2).
Third, there were 14 Implausible problems ($n = 14$, Category 3)

e.g. One pen in the stationery store costs $20. We spent $390 to buy some pens. How many pens did we buy? How much was left?

Fourth, there were 18 insufficient math problems ($n = 18$, Category 4)

e.g. Sister found a bucket of 6-piece chicken cost $199. How much did a group of 5 persons pay for a chicken meal?

In this problem, the amount of chicken bought was missing.

If the problem included, "What should they pay if they wanted to eat 3 buckets?" then the problem would be changed to category 5 (Plausible and sufficient problem).
Fifth, there were 259 sufficient problems ($n = 259$, Category 5)

e.g. A piece of chicken wing costs $35 and a bottle of Coke costs $10. The principal invites all 761 students in the school to have two pieces of chicken wing and a bottle of Coke each. How much did the principal cost?

As above, 87.2% of them were sufficient problems. During math lessons, the students made a variety of problems. The arithmetic problems included also multiplication and division problems after the third lesson. The word problems might not match real life (Greer &McCann, 1991) and they were fixed after the whole class discussions thereafter.

At home, and at the earlier stage, parent–child pairs were not used to this MPP activity, they might forget to pose math problems. Some pairs did write down responses that were not-a-problem. However, once they get used to it, they make sufficient problems. Interestingly, they remember that the provided information was

Table 2 Results on MPP instructions and children's problem posing

Math content	1. Not-a-Problem	Categories of problems (Leung & Silver, 1997)			
		2. Non-Math	3. Implausible	Plausible	
				4. Insufficient	5. Sufficient /Extraneous
Arithmetic (3 lessons)	3	0	5	5	68
Multiplication and division (4 lessons)	3	0	6	10	89
Four basic algorithms (4 lessons)	0	0	3	3	102
Total (11 lessons)	6	0	14	18	259

for posing math problems. For all 11 lessons, they did not send any non-math problems as outcomes. Finally, there was an increase in plausible problems over time (Table 2).

Attitude of Students

Measures of attitude ($n = 27$) according to three scales (Hsu, 2001) increased after MPP instruction. Scores for pretest and posttest were: Confidence in Math (mean: 13.11, 13.44); Motivation (25.93, 26.85); Communication and Interaction (mean: 25.70, 27.00). As above, the mean in the three measures scales was higher in the post-test, although no statistical differences were found in t-tests, and the number of students was only 27 (see Table 3).

Table 3 Results of math attitude test

Scale	Test	No. of students	Mean	S.D
Confidence in Math	Pretest	27	13.11	1.99
	Post Test	27	13.44	1.78
Motivation	Pretest	27	25.93	4.23
	Post Test	27	26.85	3.68
Communication and Interaction	Pretest	27	25.70	5.75
	Post Test	27	27.00	4.09

From children's interviews, children with high scores in attitude enjoyed doing math problem posing by themselves while those with low scores might prefer to just listened to the teacher. Two excerpts of each are clipped below.

High-scores in attitude:

960427S–1: I like both: "MPP and the way we had taught before. I don't have to think too much when the teacher teaches, just listen and remember. However, if I pose a problem by myself, I can pose a different or more difficult problem."

960427S–2: "All are great. But I like posing problems more. I can pose different problems. Problems in the textbook are simple and I finish quickly. Besides, I can go to the stage to present my problem when doing problem posing."

Low-scores in attitude:

960503S–5: "It is clearer when the teacher is teaching. A bit unsure about posing problem by myself and presenting on the stage..."

960503–6: "Sometimes I don't understand what my friends are talking about. Besides, I don't like to go to the stage."

4.2.2 Parent–Child Interaction and Views

Worksheets ("Let us share today's math") were provided for parents and children to do problem posing together at home (5 times). For the first three times, the teacher told parents to do revisions with children and then pose problems together. They solved the problem together. Finally, they posed one more problem to bring to school the next day. For the remaining two times, the teacher invited the child to pose and the parent to solve; then the parent to pose and then the child to solve. A total of 270 problems were posed, there were only 4 Implausible (appeared in the first and the second times), and 1 Insufficient (third time); all the remaining 265 problems were Sufficient.

Interactions of Parents and children

Mytle received the teacher's two findings.

First, all parents were serious about developing skills in calculations from the abundance of posed problems on four algorithms. They were also serious in checking mistakes in posed problems. When they found problematic problems, they interacted with children and shared results with the teacher.

Incorrect conversion rate. "*One catty is the same as 25 gm, how many gm are there in 8 catties?*" Insufficient data, "*The apartment building is 9 stories high, how many units are there altogether?*" Inconsistent units. "*One tank of water is 459 kg heavy, how many liters are in 15 tanks?*" *Inconsistent names.* The names that appeared in the word problems are not the same. (e.g. "*I bought ......*", "*How many did Tony buy?*"). They interacted with children and with the teacher until the problems were fixed.

The second was about differences in MPP between high/low scorers in math tests. The groups might returned problems with different levels of difficulty.

Single-step problems tended to be given by families with children getting low scores.

> *960427S06_P-C pose: A pen costs $500. How much did five pens cost?* This problem indicated that the low achiever child and parent posed a simple problem asking for five times as much. The working sheet showed a calculation of "500 × 5" by the child. They talked through and made up this simple word problem. Anyway, the parent's guidance was most treasured.

Complex problems tended to be given by families with children getting high scores.

> *960427S22_P-C pose: 2 sets of furniture cost $5,600, 4 refrigerators cost $8,820, 3 air conditioners cost $4,998, and 5 fans cost $3,800. Dad bought 2 of each, which one of them paid the most, and what was the difference between paying the most and paying the least, and what was the total amount of money he paid?* This problem indicated that the high achiever child and parent posed a set of connected problems. The pair asked for the total amount, most expensive, least expensive, and the difference.

Mytle reminded the teacher that the focus was on parental interaction and not on differences in math performance. However, the teacher noted individual differences for a reason: to do future follow up sessions when in need.

Views of Parents and children

Parents' views on problem posing.

The teacher interviewed 6 parents: They expressed that normally they read quickly and signed the teacher-parent communication book. The time they spend ranged from 30 min to an hour. After problem posing was introduced, they spent more time on children's homework, at least 10 to 30 min more. Besides additional amount of time used (quantitatively), they discussed with their children and began to understand children's thoughts (qualitatively). Parents expressed that they learned together with their children.

> *"It is good that Sandy posed and solved problems at home, she tried to think hard and be smart."*
>
> *"I become co-learner of David, we make up problems together, we solve problems together. It's a long time since I was in elementary school and there are new ways to do math now."*
>
> *"I learn more on how to supervise Jenny's homework after this problem posing exercise"*
>
> *"I begin to realize the changes in curriculum and from now I know better how to explain to Alson and trap his math thinking..."*

Children's views on problem posing.
The teacher interviewed 6 students: Three of them normally did homework with their mothers, one with their father, one with their elder sister and the last one with grandparents.

> *"I do homework by myself until mom did problem posing at home"*
>
> *"In doing problem posing at home I can revise math lessons"*
>
> *"Dad and Mom are usually busy. They just sign the communication book without reading my work. After introducing problem posing at home, you guess what happened? We spent more time doing math homework together, at least 10 to 30 minutes more than before."*

4.3 *Mytle's Discussion on Two Cases*

Parents are resources. It is feasible for teachers to use written tasks to involve parents to do problem posing and solving at home not just for grade one (Leung & Wu, 2000) but from grade one to grade four. Parents provided useful information about children. The result is consistent with a prior study showing that parents could be resources (Civil & Bernier, 2006). In addition, the returned work revealed the perspectives, feelings, and activities of the members of the families. For example, for grade one, the parent–child might use counting to solve problems for the amount of money for a few cans of alcohol. They did not use multiplication or addition. When dyads from the third grade wrote After-dinner-math, it happened to be the Lunar New Year, the word problems were mostly related to the Chinese New Year. In real life, after counting lucky money from red packets, children learned to do money conversion in order to find the total amount received during the Chinese New Year. They did not need to wait til grade three. In school, children came across problems with money conversion money in grade three.

MPP and MPS make it interesting and meaningful. As Williams and Williams (2021) reported, when homework was interesting and designed to cater to the needs of families, children and parents loved homework. In this study, teachers' design of MPP enabled dyads to pose and solve a variety of problems together at home. Problems created by children together with parents were different from those in textbooks and were more diverse. Usually, problems given by textbooks or the teacher were related to a single unit (e.g., coins, paper money, money conversion) and were presented according to the unit. However, problems posed by children and parents might not be related to the unit on money (e.g., "*Which coin is heavier?*" where the unit is weights). The problems they posed ranged from one strand to another. In addition, children were influenced by their parents' work. Through the process of formulating problems with adults, children were highly motivated. With excitement, they were eager to solve problems they created together, especially real-life related problems.

Parents and children found solutions using ways that make sense to children. When children worked together with adults to formulate problems, the adults could arouse children's motivation to solve problems and worked harder than by themselves. This did not mean children learned harder math or learned earlier from their parents. Mytle shared what she knew and reminded parents that the simple division method was not taught until the second semester of grade two. Therefore, children and parents solved them using their own ways. In the need to find how many ice creams can be bought by $24 if each costs $8, the grade one child did not use "24 divided by 8", but they could find "3" as the answer using counting. "$24, $16, $8, three ice-creams". Parents told the teacher that they used buttons to help the child. Mytle also noted that by grade 3, students' language proficiency greatly improved.

Unlike case one, the math teacher in case two taught MPP and then follow up with MPP at home. In the sheet, students were reminded to share with parents what happened during math class. Hence, parents and children posed very challenging problems and solved them. When referring to the attitude of students, high students considered problem posing as important in getting themselves to do the math themselves instead of just listening. However, students with low scores prefer to be sure and want to just follow the teacher. This finding on individual differences in posing and then solving problems is an implication for teachers. To start, let the low-score students follow and have a sense of nice feeling of accuracy. Once they master math facts and concepts, the teacher can encourage students to take charge of their own learning in posing and solving. Patience is needed.

5 Conclusion and Implications

Mytle's initial attempt to have parents involved in children's math learning using simple written tasks was reported. In both cases, it was feasible to use written tasks to motivate problem posing during class and after school. The actions taken were responding to the international emphasis on parental involvement (OECD, 2020) and in the country's national curriculum documents (MOE, 2014) and particularly on math problem posing (MOE, 2018). In this chapter, results on both cases were positive. For case two, the performance on MPP of children and parents improved over time; and, children's attitudes also changed over time. Views from parents and children were positive towards problem posing instruction and problem posing at home. Also, MPP at home increased interactions with family members. Siblings were also invited to do this activity. The activity increased parents' awareness of children's learning in school. The teachers involved found that their parents spent more time communicating with him. It is concluded that problem posing in class and at home can enhance interactions among parents, children, and teachers. However, there are limitations on using written tasks. It is recommended that further studies on parents can be carried out, not just on completing written work at home. For example, the exercise of reading results on posing word problems and then categorizing problems might be an interesting activity for future parents' workshops. Parents from other grade levels can be involved (kindergarten, grade five, and, grade six). Indeed, more effort is needed for future studies to study parents, who were involved as resources in children's math learning (Civil & Bernier, 2006).

Mytle was on the way to prepare an agenda for actions on enhancing parental involvement.

References

Barrett, J., & Smith, J. (2017). The learning and teaching of measurement: Coordinating quantity and number. In J. Cai (Ed.), *Compendium for research in mathematics education* (pp. 355–385). National Council of Teachers of Mathematics.

Civil, M., & Bernier, E. (2006). Exploring images of parental participation in mathematics education: Challenges and possibilities. *Mathematical Thinking and Learning, 8*(3), 309–330.

Creswell, J. W. (2009). *Research design: Qualitative, quantitative, and mixed methods approach.* Sage.

English, L. D. (1998). Children's problem-posing within formal and Informal contexts. *Journal for Research in Mathematics Education, 29*(1), 83–106.

Greer, G. B., & McCann, M. (1991). Children's word problems matching multiplication and division calculations. In F. Furinghetti (Ed.), *Proceedings of the fifteen international conference for the psychology of mathematics* (Vol. 2, pp. 80–87). Author.

Heller, J. I., & Greeno, J. G. (1978). *Semantic processing in arithmetic word problem-solving.* Paper presented at the Midwestern Psychological Association Convention.

Hsu, H. Y. (2001). *Experiment design of portfolios assessment and paper-and-pen test to the fourth-grade students of elementary school on Mathematics Concept, mathematics communicating capability, and mathematics learning attitude.* Unpublished master thesis. Institute of Education, National Chiayi University.

Leung, S. S. (2008). *Math wonder bag: Get out! Math sleepy bug.* Outside The Box (OSB) Publishing Co., Ltd. (In Chinese, 174 pages).

Leung, S. S. (2013). Teachers implementing mathematical problem posing in the classroom: Challenges and strategies. *Educational Studies in Mathematics, 83*(1), 103–116.

Leung, S. S. (2015). *Math after dinner: enjoy family math time.* Outside The Box (OSB) Publishing Co., Ltd. (In Chinese, 140 pages).

Leung, S. S., & Silver, E. A. (1997). The role of task format, mathematics knowledge, and creative thinking on the arithmetic problem posing of prospective elementary school teachers. *Mathematics Education Research Journal, 9*(1), 5–24.

Leung, S. S., & Wu, R. (2000). Sharing problem posing at home through diary writing: Case of grade one. *The Australian Mathematics Primary Classroom, 5*(1), 28–32.
Leung, S. S., Wu, P. J., & Chen, C. H. (2012). A study on parents attending math study group designed by grade one elementary school class teacher. In T. Y. Tso (Ed.), *Proceedings of the 36th conference of the international group for the psychology of mathematics education* (Vol. 4–293). PME.
Ministry of Education of Taiwan. (2014). *Curriculum guidelines of 12-year basic education Gineral guidelines* (in Chinese).
Ministry of Education of Taiwan. (2018). *Curriculum guidelines for the 12-year basic education elementary school, junior high school, and upper secondary school-the domain of mathematics* (in Chinese).
National Academy for Education Research. (2020). *Math Curriculum Guidebook* (in Chinese).
Nesher, P., Greeno, J. G., & Riley, M. S. (1982). The development of semantic categories for addition and subtraction. *Educational Studies in Mathematics, 13*, 373–394.
OECD. (2020). Parental involvement in school activities. In *PISA 2018 Re-sults (Volume III): What School Life Means for Students' Lives*, OECD Publishing, Paris. https://doi.org/10.1787/29fed428-en
Polya, G. (1954). *Mathematics and plausible reasoning* (Vol. 1–2). Princeton University Press.
Smith, J. P., & Barrett, J. E. (2017). Learning and teaching measurement: Coordinating quantity and number. In J. Cai (Ed.), *Compendium for research in mathematics education* (pp. 355–385). Reston.
van den Heuvel-Panhuizen, M., Middleton, J. A., & Streefland, L. (1995). Student-generated problems: Easy and difficult problems on percentage. *For the Learning of Mathematics, 15*(3), 21–27.
van den Heuvel-Panhuizen, M., & Buys, K. (2008). *Young children learn measurement and geometry*. Sense.
Williams, K., & Williams, H. (2021). Mathematics problem-solving homework as a conduit for parental involvement in learning. *Evaluation of a Pilot Study, Educational Review, 73*(2), 209–228. https://doi.org/10.1080/00131911.2019.1566210

Chapter 12
Roles of Parents When Involved in Workshops for Children's Math Problem Posing and Solving: Examples From Grade Five and Grade Six

1 Introduction

This study is part of a larger project on parental involvement in children's mathematics learning in Taiwan. In an elementary school within walking distance from a university and near a subway station, a mathematics teacher educator (MTE), Mytle, offered bi-weekly Friday mathematics camps for parents and children year-round over time. Parents from this school shared the news with other parents; and later on, participants from other schools joined. Preliminary results were given by Leung (2012) and in a book (Traditional Characters in Leung, 2015; Simplified Characters in Leung, 2019). With funding from MOST of Taiwan, this project for lower grades extended to middle and upper grades and included kindergarten (K-6, see Leung, 2018). In order to promote interactions between adults and children, research-based math tasks were carefully developed so that children could not complete them in the absence of at least a parent. The tasks were called *Family Math Tasks* (FMT) and included both the family and math dimensions (see Leung, 2023). Leung (2023) included characteristics and examples of FMT for young children and how parents used strategies to enact them.

Why Parental Involvement?
Education for the next generation means a responsibility for all. "Education" is spelled with all vowels (a, e, i, o, u). When educators read aloud "education", they voice the exchange of ideas across nations and promote education for all ages. In Chinese, education is "jiàoyù 教育" which means "to cultivate for growth". At home, jiàoyù begins; children's parents were their first teachers. Research on large-scale comparative studies confirmed the importance of parents' involvement (e.g., Chiu & Zeng, 2008). It was found that parents contributed to academic achievement, positive behaviors, and, emotional development of children (e.g., Stevenson & Lee, 1990). Hence, parental involvement in children's education is gaining increasing attention in many countries.

S. S. Leung, *Mathematics Problem Posing in Action*,
https://doi.org/10.1007/978-981-96-8480-9_12

Examples of projects on home-school partnerships were reporting results from around the world. They are project NNPS (National Network of Partnership Schools) in the US (Epstein, 1995); for families of high socioeconomic status in a primary school in New Zealand (Maher, 2007); for Latino families in project MAPPS (Math and Parent Partnerships in the Southwest, see Civil & Bernier, 2006), home school partnership in Hong Kong (Ho & Kwong, 2013) and SEND (special educational needs and disabilities) project in UK (Broomhead, 2019). These projects are exemplars showing successful learning outcomes by joining combined efforts from parents, teachers, and teacher educators.

When the subject area is mathematics, what can parents do to be involved in their children's education? When children are not young children and grow to upper grades (grade 5, grade 6), what can parents do to continue paying attention to children's learning? How can parental involvement constitute knowledge growth for both parents and their children? This chapter reports on a case where Mytle, a math teacher educator, used math problem posing (MPP) as a means to foster parental involvement in children's mathematics learning.

The research questions are:

1. What roles did parents assume as they attended this Family Math Camp over time?
2. What exemplars of parents using MPP, can be collected as evidence of roles that carry the potential to enhance the learning of parents (and children)?

In the larger project, parents with children from kindergarten to grade six participated. Here in this chapter, results were taken from analyses of data on parents for grades 5 and grade 6 (the last two grades of an elementary school in Taiwan). Then Chapter 13 (Chapter 14) will include analyses of data on parents for grade 3 and grade 4 (grade 1 and grade 2).

2 Background Literature for Action

2.1 Curriculum Standards Emphasize Parental Involvement in Mathematics Learning

In both the US and Taiwan, parental concern for children's learning of mathematics is explicitly emphasized in curriculum standards documents. In the United States, "*Families become advocates for education standards when they understand the importance of high-quality mathematics education for their children*" (NCTM, 2000, p. 378). In Taiwan mathematics curriculum standards documents, there was a reminder for parents to be positive and not to push children to overwork. "*Learning math should be a happy experience for students. When parents find children having a hard time with learning mathematics or having low scores there is no need to be anxious*" (MOE, 2014). In this document, parents are reminded to assure students, to encourage them to complete homework and not to give up learning math.

Parents' attention and family education appeared in the latest two curriculum standards. The 9-year and 12-year curricula are the two major education innovations in the past 20 years in Taiwan with the latter extending from the former to 12 years for addressing justice (MOE, 2014). The document listed a total of 19 issue-based curriculum integration to languages, math, science, technology, and arts. Family education is one of the 19 issues.

Research studies on parental involvement included activities and strategies for enhancing mathematics learning that were useful in intervention programs for parents. The activity can be in the form of games (Leung & Lo, 2010; van den Heuvel Panhuizen & Buys, 2008), reading a picture book (van den Heuvel Panhuizen et al., 2009), posing problems and writing diaries (Leung & Wu, 2000), or, completing a math trail (English, Humble & Burmes, 2010). In an intervention project on parental involvement (Civil & Bernier, 2006) the investigators provided opportunities for parents to be involved with handling mathematics topics (like proportional thinking) and cope with family interactions around mathematics (techniques in asking questions). The reported stories explained that the work with parents was challenging but yet not entirely impossible. On top of that, empowering parents contributed to growth for them to lead other parents. When the focus is on mathematics problem posing (MPP) as parental involvement, what results can be obtained? It was worth studying how adults assume their roles in involvement, including instances where they might not be able to solve the problems their children ask for help.

2.2 *Combined Efforts: Teacher Educator, Teachers, and Parents as Co-Learners*

The investigator used a framework by Jaworski (2008), explained in a Venn diagram (Fig. 1). It represents how a mathematics teacher educator (MTE) and teachers share knowledge. One circle carries MTE's knowledge of research and theory while the other circle holds teachers' knowledge of students and schools. The shared knowledge is in the intersection of the two circles; the passing of knowledge is bi-directional. In this study, a third circle represents parents. Parents share knowledge of children that the teacher and/or teacher educator might not know. In Fig. 1, adults who contribute knowledge and learn from others are named "co-learners" (Jaworski, 2008).

On the same plane. When two circles overlap to result in an intersection, they must lie on the same plane. If not, there is no intersection even though they appear to overlap when viewed from above or below (Fig. 1). If any of the circles are not on the same plane, there is no intersection. When there is no sharing of knowledge, the above co-learning model will not work.

If a teacher is staying quiet and just listening to the teacher educator, then the sharing is not likely to happen. The most likely scenario is to result in the teacher

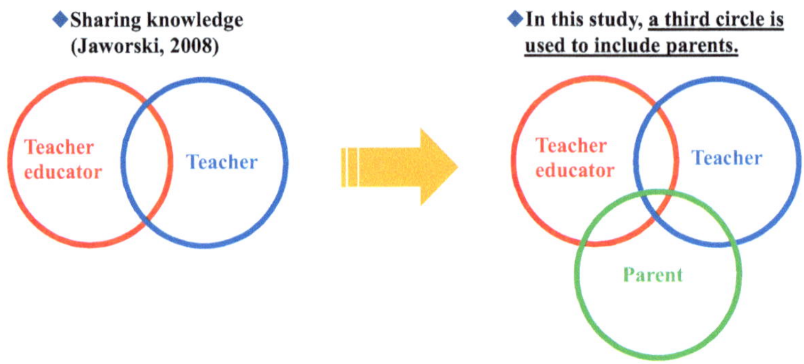

Fig. 1 Co-learners lying on the same plane

educator teaching teachers (one-directional). To avoid this scenario from happening, both teacher educators and teachers can work on something new for both. For example, integrating new ideas into curriculum standards (Leung, 2013). The same applied to three intersecting circles. The three parties can work on a topic that is unknown to all three, such as "What can parents do to motivate children to pose and solve math problems after school?".

2.3 Parents' Role When They Work with Teachers and Teacher Educator

What roles do parents play in school? Volunteers for gardening, arranging books in the library, and helping out with coffee for meetings. Even in the classroom, they are helping with notice boards but very rarely play the role of resources in teaching. Parents often receive volunteer work from teachers. In that case, the circle representing parents is not lying on the same plane as the circle representing teachers. Under what conditions will parents share knowledge on children's learning? How can they be resources? What roles can they take as co-learners of teachers and teacher educators? It is worth studying how that can happen.

Investigating parental roles is important to empower parents to be active enough to take the lead in sharing responsibilities for children's learning. Epstein (1995) identified six types of parental involvement: parenting, learning at home, communicating, volunteering, decision-making, and collaborating with the community. Among the six, anxious parents tend to emphasize children's learning at home and teach children at home. Indeed, "parents as teachers" is one role identified by Civil and Bernier (2006); the other roles were learners and facilitators. It is worth studying in Taiwan, where parents are usually anxious about children's learning, what roles will parents assume?

3 Method

The case, participants, and materials will be given below. A more detailed description of the setting of the family math camp and family math tasks (for young children) was given in Leung (2023).

3.1 Case and Participants

This is a case study and the method follows that of Yin (1994). The research team consisted of a mathematics teacher educator (Mytle, also the author) and her research team. Participants also included elementary school teachers, parents, and children. The description of Mytle as a math teacher educator, teacher, and parent is also given in Chapter 1.

A Typical Friday Math Camp (Time: 13:40 to 15:20)

There were two 40-min sessions and a break. During the first session, parents attended when the teacher educator explained how children learned mathematics; conducted focus group interviews, asked parents to fill in a questionnaire, or led a discussion on "What if my child…". While parents were having class, their children were in a separate classroom making up teaching aids required for playing games with parents. For example, shading colors on fraction cards and cutting them into a deck of poker cards for playing equivalent fractions. During the second session, parents and children attended class together. The teacher educator introduced mathematics games to be extended at home. The seating arrangement is "one table, one pair". Mytle made sure each pair could play on their own before the whole class played together. Finally, she sent out a follow-up activity and a diary sheet for parents to complete at home and to share what happened during the next Friday meeting.

The original arrangement, where both parents and children came together during the Friday afternoons, was for all grades. Later, Mytle found out that children in grade 5 and grade 6 had classes on Friday afternoons and could not join parents. However, parents expressed interest in participating even without their children coming along. Therefore, parents and adults worked in pairs and asked one of them to play the role of a child to complete the math activity, and once they managed, they exchanged roles. They took turns being "children" in the parent–child pairs. After the camp, they tried the math activity with their children at home and shared the results during subsequent meetings.

3.2 Materials with Potential for Problem Posing

By referring to Civil and Bernier (2006), Mytle developed tasks, including problem posing, where parent-parent pairs learned mathematics together or attended to how their children learned mathematics. Later, these participants also took the initiative in deciding what to do, but not until after two months. Note that problem posing, the theme of this book, was not singled out as a compulsory/separate activity in this Friday's math camp.

These activities were in the form of a game, reading a picture book, posing problems, paper folding, writing a diary, organizing a fun fair, or completing a mathematics trial. There were two characteristics of each activity: adult pairs (parent and "child") learned some mathematics together, and, the parent went home and started to interact with the child to complete the math activity they tried during the math camp (i.e., could not be completed by just one person alone). To explain how to motivate parents and children to learn mathematics concepts and interact with each other, one example task in the Shape-and-Space Strand (S) is given here.

Example Task: STORYING with tangrams (Space Strand). For plane figures, the child and the parent each do paper folding which resulted in paper Tangrams. Each person in the family used the seven pieces to create an object (e.g., a tree; or a kite). When asked how to make shapes, Mytle did not give examples for fear that they just followed exactly the same as her example. Instead, she gave examples of "don't do". An example of "don't" is "the pieces could not overlap nor separate". Another example was, "Don't miss out on any piece, use all seven pieces in the Tangram". Mytle believed that an example of "do" (e.g., a cat) might attract imitation (see a prior project on teacher development in Leung, 2013). After each of the Parent-"child" pairs finished making a shape (e.g.: a kite and a tree) the parent and child made a story using a tree and a kite and recorded the story in written form. Since the written form was for sharing, Mytle told participants that they are called "math diaries". Later on, family members used the story to ask and answer questions. Sharing and communication were emphasized and spatial sense was enhanced. Later, they could remove the pieces, make other shapes, and write different stories. With another parent, siblings, and grandparents more shapes were made and a longer story could be made. This family task crossed two dimensions: Family dimension meant members' interaction, and math meant making out geometric shapes using common sides and considering areas.

Although the activities were developed by the teacher educator, she sent them to three elementary school teachers to review. Revisions were made until all agreed on the timing, appropriateness of materials, and potential opportunities for interaction of the child with parents.

3.3 Data Source and Analyses

In each school year, there was a total of 16 bi-weekly Friday Mathematics Camps for parents. However, the persons from families who attended varied according to grade. Data sources were: questionnaires, diaries, parents' focus group interviews, worksheets, and videos; the data were analyzed qualitatively as in Creswell (2009). Data were coded by two independent raters and reliability was checked. Disagreements were discussed and resolved through e-mails or telephone discussions.

4 Results

Results will be reported using examples of math activities (Friday math camps) used to provide affordance for parental involvement. These activities were developed and announced as a final module for parents to register to come.

Module for parents. Mytle referred to curriculum standards and textbooks series to develop research-based activities suitable for parent-parent Math Camp during the bi-weekly Friday afternoons. There was a total of 4 textbook publishers. She read the chapters and the school calendars of schools and purposely selected two lessons that were commonly taught by all four schools over the two weeks. "*My intention was for family interactions and not on math performance but I include mathematics contents in the module to make it more attractive to parents.*"

In the module, the catchy names are BIG TWO on factors/multiples, fraction UNO, CANDY with different surface areas but unchanged volume, SEPARATED sectors, TRADE with decimals, STORYING with tangrams, FANCY symmetry, MOUTHFUL of milk in c.c., M&M bar charts, and RECYCLE equation cards. The module included names of math contents and were embedded in real-life applications. Parents who attended the Friday sessions were engaged in activities but not lectures. The intended objectives were math contents and children's thinking, and, techniques in family interaction in a mathematics context. Mathematics contents for parents were: multiplication and division, factors and multiples, sectors of a circle, equivalent fractions, computation with fractions (decimals), surface areas, units in time and conversion, volume and capacity, symmetry, unknown and equations, ratio, rate and proportion, percent, tangrams, cube and cuboid, H.C. F. and L.C.M, bar charts, straight line graphs and pie chart, speed, and, circumference and area of a circle. Children's thinking for parents were: Semantics in word problems from textbooks, knowledge of variables, van Hiele levels, and, how to "lie" with statistics, children's work from textbooks, examination papers, and manuals for manipulatives.

The group also shared findings of parents on children's thinking and adults' attempts and evaluation of strategies. Parents posed problems of their own, with peers (other parents), and their family members. They reported orally over the meetings or in the form of written form called, "math diaries". One focus group meeting was held at the end of the term.

Over the two years, there were instances of multiple roles of parents over time. The following were cases when parents participated and played the roles of *learners, partners/opponents, teachers/observers, facilitators, and leaders*. Below are results on multiple roles with exemplars that carry the potential to enhance the learning of parents (and maybe, children).

4.1 Parents as Learners

First, parents learned mathematics. For example, they learned about factors and multiples. In the game, BIG TWO, the player had to display one number card and its factors. Parents learned how to find factor cards and multiple cards. They remarked that they were "grade 5 students". *"I almost forgot what they [Factors and Multiples] are and their relationship until I played this card game"* (2017.11.06). Interestingly, after they followed the rules to play this card game, parents wondered if they could use the same deck of cards to play other games. They used problem posing strategies such as "what if" and "what if not" (Brown & Walther, 1983) to create new games. They posed problems using "Can we change how we use the deck of cards?", "What if we add or remove cards?", and, "What if we change the regulations on how to win". Another example was about the unit on the "surface area of a cube", they cut a paper cube and obtained a configuration of 6 joint squares. They realized that the surface area was 6 times the area of one face. Then they found in reversal, that when they folded the 6 joint-squares surface they could wrap a cube. Mytle asked them to pin up to the board and to observe results from other groups. They were surprised to see that the configurations of 6 joint squares were not necessarily the same. Mytle used a set of 6 magnetic cube rings to let them try. "*Please try to make as many different configurations as you can.*" Each group put up the figure on the blackboard once they found a new one. Finally, pairs of parents learned that there was a total of 11 different configurations of 6 joint squares that could make a cube (2018.02.18).

Second, parents learned about children's math thinking. For example, the teacher educator sorted children's answers and invited parents to review her collection of typical work. Parents talked to each other; they guessed how their children would solve problems and suggested strategies and misconceptions. During the "Q and A" session, parents might also ask questions such as why some of the answers were wrong. Mytle contrasted them to the correct methods. With confidence, parents even came to the blackboard and tried to answer in the way they guessed their children would answer.

In one exam paper on fractions subtraction on mixed fractions (a whole number together with a fraction), Mytle knew that children tended to forget to reduce the larger number by "1" after borrowing "1" to subtract the fraction part. A parent guessed how her child did and attempted to check the work of her child when she returned home. She was correct and shared during the subsequent meeting. "*If the fraction part of the larger mixed fraction is MUCH larger then my son would not make this mistake*". "*I posed one problem for my son to do again to check if my guess was correct. Guess what? I was correct. It was such an exciting thing to share today...*" (2017.11.20). Her problem was "$4\frac{2}{5} - \frac{1}{20}$". She was one of those parent learners who reviewed, guessed, posed extra problems for children to try, and checked her guesses.

Third, parents learn how to respond to children. One day, parents came to the meeting excitedly and shared the effects of different ways to talk to children. Parents and the teacher educator shared how to talk to children "*I pretended not able to solve the problem and ask my girl to teach me* (2017.12.29)", "*At home when I talk to Betty about math, I would not use the word 'you'; I use 'we'...my girl worked promptly when I changed the 'you' to 'we'...*" Finally, they shared how to use gestures and when to use praise. "*When Tim asked me math problems that I think he should be able to try himself I covered my pair of ears with hands* (2017.12.15) *and did not help*; *and when he attempted, I gave him a BIG thumb up as praise, no matter whether he succeeded or not*".

4.2 Parents as Teachers (Observers)

First, parents make up problems for children to solve. During meetings, parents taught or observed how to teach. Parents were solving problems and also posing problems for other parents (playing the roles of children) to solve. When they went home, they shared posed problems from the group with children and parent–child solved problems at home. For example, they posed problems for children to solve with manipulatives. In the activity (GUESS Me) the mathematics material was unknown (Algebra Strand) one of them used plastic chips as unknown and posed this problem. "*In a flower shop, the cost of 4 flowers is $12, when one purchases 3 flowers, how many dollars should one pay?* (2018.05.25)".

Second, parents respect children's thinking. Mytle shared changes in the math curriculum that suggested respecting children's thinking. As long as the solution was reasonable the teacher would mark it as correct. There was no need to insist on a reproduction of the same solution printed in the textbook. Parents compared children's responses as if they were teachers. "*I realized the way my boy learned mathematics was different from mine. My husband now allows our two children to use their ways to represent solutions. I learn to respect them. Now when we have lunch over the weekend, I let my kids decide what they like to eat, as long as the nutrition is there...* (2018.01.12)" Parent gradually practices respecting children's thinking for math solutions and other instances on decision making in real life.

Third, parents provide resources. Over time, Mytle witnessed parents do the monitoring, provided resources, and advised on math content only when in need. How did family members play "RECYCLE equation cards" at home?

A set of cards each with either "0" to "9", one of the four operations symbols, equal sign, or "………" meaning remainder. The goal is to use as many cards as possible to form correct expressions. The winner is the player who uses the most number of cards.

"When we played "RECYCLE equation cards" I let my two children flip the cards out. The older one asked if he could use division so that he could use more cards but his younger brother (who only knew addition/subtraction) did not understand what division was. Until then I just monitored and did not participate. After "division" was used by the elder brother, I posed a problem for the younger one, to check by using subtraction." Diaries (5–1; 2017.10.06; Parent: HC Children: YC)

4.3 Parents as Facilitators

As time went by parents were familiar with each other and made friends. Mytle started to provide opportunities for two parents to be facilitators. Her reason is, "*If I ask just one to come to the blackboard to share thoughts then she or he may be too shy and say no. Usually, I invited two at a time….*" In Taiwanese culture, it is true and typical of a parent to humbly say "*I am not the next person*" or "*Please invite xxx instead*" even though she or he can. During the first attempt, Mytle successfully invited two parents, and they boldly shared thoughts and attempts at home. Three attempts are given below.

Symmetry. Mytle invited a parent to present diaries on symmetry and do storytelling about what happened at home. After doing so other parents would join in. "*It was not until I was asked to take a turn to tell during the next meeting then I made an effort to record down what my son did*". Another parent said, "*I was asking Tom what symmetry was. Instead of answering my question verbally he folded a paper and cut out a yellow butterfly which is symmetric. Next, he used grid paper and made 2 more figures by line symmetry in red (left) and in blue (right)...*" (see Fig. 2).

Surface area of a solid. More parents joined as facilitators afterward. One parent told other parents that she used cubes to make a solid and then posed a problem for her child to find surface area. The child did counting the top and the bottom (9, 9); then the left and the right (9, 9), and, finally the front and the back (9, 9). The child then added to "54" to get the total number of faces. (see Fig. 3).

Percent on grid paper. Parent TY facilitated a discussion as she shared results on percentages with her girl. Parent TY used 10 × 10 grid (100 squares) paper to play a game on area by percent with her girl PJ. The game is "*FANCY percents, I Draw What*

Fig. 2 Worksheets (5–2; 2018.04.27) Parent: YW; Child: HC

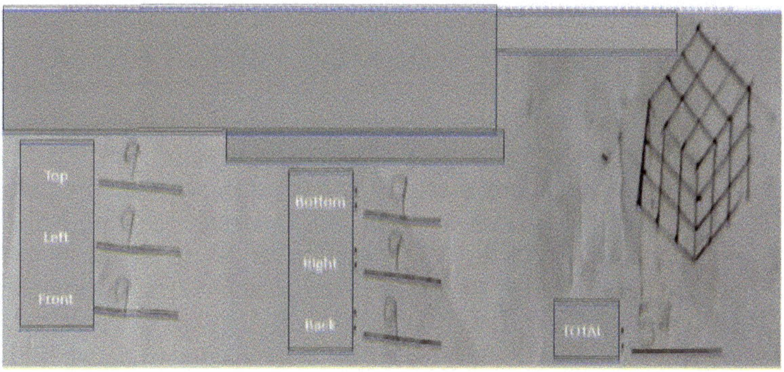

Fig. 3 Worksheets (5–2; 2018.05.04): "Finding surface area"

You Say". One said "5%" and the other shaded 5 squares. What followed was 20%; 12%, and 35%. (see Fig. 4: First row, from left to right). When the parent posed the subsequent problems as "40%" and "38%" the shaded area given by her girl were two Chinese characters (see Fig. 4: Second row, left-1 and left-2). Mother and child then changed the game. This time, one player posed the problem with a drawing and the other solved the problem by telling how many percent. The house was 43% while the fish was 26% (see Fig. 4: Second row, left-3 and left-4). Suddenly, the child wrote her Chinese name and wrote three different percentages on the characters respectively (see Fig. 4: Third row). To end, she drew "*I* ❤ *U M A*" and wrote 18%, 35%, and 34%, then 35% and 30% next to I, ❤, U, M, and A (see Fig. 4: Fourth row and fifth row); representing the percentage of area shaded. She signed using a smiling face (39%). Indeed, "*I love you ma*" and a signature (smiling face) was a very sweet ending (see Fig. 4 for shadings and percentages).

Fig. 4 Worksheets (5–2; 2018.06.08): "I Draw What You Say" (Percent) Parent: TY; Child: PJ

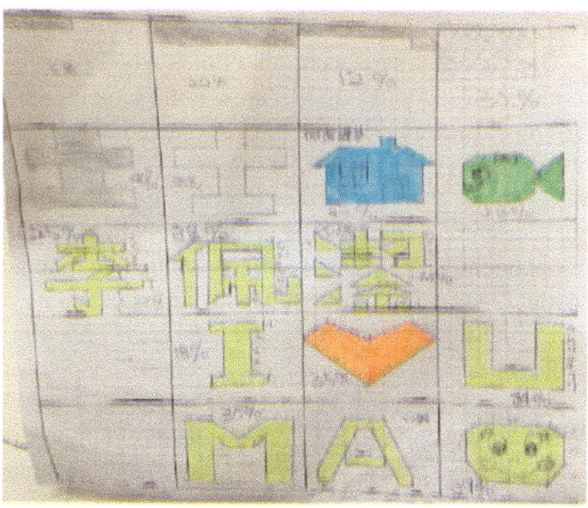

Parents interacted with other parents and helped when other parents were in need. They checked each other's work before reporting findings to the group. The first example was parents folding paper into a tangram, cut into seven pieces then using them to make shapes and name them. Finally, they used the named shapes to create a story and came to the stage to tell the story. *"The presentation was most unforgettable. Our family members seldom had a chance to present results together. Indeed, we realized that the family time was so much fun by just sharing"* (Diaries, 2018.01.02). The second example was on "RECYCLE equation cards". To play, display a set of cards each with either "0" to "9", one of the four operations symbols, equal sign, or "........." meaning remainder. The goal was to use as many cards as possible to form correct expressions. The player who used out the most was the winner. A daddy and a grandma sat together and did not know how to start. Then the grandma remembered what her grandchild challenged her after school. She started. She posed a problem by using seven cards ("0", "1", "2", "5, "6", and " = , + , -, x, ÷ ") to ask the dad to make one equation and tried to use as many cards as he could. The daddy when challenged by the grandma, and when pressed with time, finally made "$12 \times 5 = 60$". Finally, both of them managed to use many cards. They came first among the pairs in the class.

4.4 Parents as Players (Partners or Opponents of the Child)

Partners. When the math activity was on addition using FractionPOKER cards, the circles were drawn and players used coloring to represent the fractional parts. Parent and child worked towards a common goal–two fractions adding to "1". They were partners who collaborated. "*When we played fraction poker cards, we reminded each other when the shaded parts of two cards are added to one circle then the two fractions would add to "1". My girl and I found this game fascinating*". Diaries (5–1; 2017.10.27; Parent: CY Child: MY).

Opponents. There were also times when parents and children were opponents in playing FractionPOKER cards. For example, when the game was on equivalent fractions. Each player took 5 cards, picked one from the other player removed pairs of cards with equivalent fractions then refilled the hand to keep having 5 cards after each round. The rule made each player try to win the other by getting more pairs but save cards to prevent the other player from winning. No matter whether partners or opponents, parents, and children expressed that they love the family math task of Mytle. One family wrote, "*When we played this card game our eyes watched each other and we were so close to each other, eyes and hearts, body and soul, it was so warm and lovely!*" Finally, a parent responded loudly in a meeting, "*The secret in playing games with them[son/daughter] is–NOT TO WIN*" (2019.02.22). Parents found that when they were opponents of children, they better not win the game.

Winner or Loser. After trying Mytle's game on FractionPOKER, one parent reported using math problem posing (MPP) to play this game. "*We draw two cards from the deck of cards and pose a problem using the two cards.*" For example, they asked which of the two fractions is greater. Find the sum of these two fractions. From the math diaries, a mother wrote "*On the next day only two of us played the game, Dad said he was busy. Daisy (my girl) added that the one who won must hold the loser warmly to console the loser...so during the next game, I purposely lost the game. Daisy held me in her arms...that was so sweet...Guess what? Her dad, upon knowing this part of the game, was willing to participate and let Daisy win for the rest of the game.*"

As above, the investigator identified multiple roles of parents over these two years. The first three were also given in Civil and Bernier (2006). Similarly, the above roles were also found in other grades (grade one through grade four, K1& K2); using other family math tasks given by Mytle with relevant contents from the math curriculum documents and textbooks.

5 Conclusion

In a prior study on parental involvement, Civil and Quintos (2023) organized parent workshops for bilingual families and studied how parents and children do mathematics together. The workshops built on two related theoretical concepts: funds of knowledge and parents as intellectual resources. Their results indicated that mothers and children positioned themselves and learned together as equal partners. In this study, and through one particular lens this time, "problem posing". Results on the analyses of parental multiple roles and examples at home were reported. The other potentials of learning mathematics and parent–child interactions via problem posing or other family math games in the workshops will be left to the imagination of the readers.

In the model given earlier (Jaworski, 2008) the intersection region of three overlapping circles is the key to success. Teachers, parents and MTE all shared knowledge about children. The findings indicated that when working with parents, it is feasible to engage parents to take multiple roles. Mytle and their teachers found they learned a lot from parents while simultaneously sharing children's math thinking to parents. Parents concluded that "*A task was best when it provided an opportunity to learn mathematics and as well enhance interactions of family members.*" (i.e. the family and math components of a family-math-task, see FMT in Leung, 2023). Results indicated that research-based family math tasks engaged parents as learners, teachers/observers, facilitators, and, players/opponents. Such roles may also lead to the potential for parents to take charge of children's math learning in school, at home, and in the community.

This chapter closes with anticipation of the next step: to empower parents, to include them as exemplars; and, to assist them in leading other parents in the community. Civil and Bernier (2006) reminded parents were intellectual resources; the resources were not limited to helping out in the cafeteria or doing notice boards. Our findings also support that the resources can be an enhancement not just in learning for both parent and child but also for a more intimate relationship in the family. More examples of parents' roles and exemplars will be given in the next two chapters.

References

Broomhead, K. E. (2019). Perceived responsibility for developing and maintaining home-school partnerships: the experiences of parents and practitioners. *British Journal of Special Education, 45*(4), 435–453.

Brown, S. I., & Walter, M. I. (1983). *The art of problem posing.* Franklin Institute Press.

Chiu, M. M., & Zeng, X. (2008). Family and motivation effects on mathematics achievement. *Learning and Instruction, 18*, 321–336.

Civil, M., & Bernier, E. (2006). Exploring images of parental participation in mathematics education: Challenges and possibilities. *Mathematical Thinking and Learning, 8*(3), 309–330.

Civil, M., & Quintos, B. (2023). Mothers and children doing mathematics together: Implications for teacher learning. *Teachers College Record, 124*(5), 13–29.

Creswell, J. W. (2009). *Research design: Qualitative, quantitative, and mixed methods approaches.* Sage.

English, L. D., Humble, S., & Burnes, V. (2010). Trailblazers. *Teaching Children Mathematics, 16*(7), 402–409.

Epstein, J. L. (1995). School/family/community partnerships: Caring for the children we share. *Phi Delta Kappan, 77*(9), 701–712.

Ho, E. S. C., & Kwong, W. M. (2013). *Parental involvement in children's education: What works in Hong Kong.* Springer.

Jaworski, B. (2008). Mathematics teacher educator learning and development. In B. Jaworski & T. Wood (Eds.), *The mathematics teacher educator as a developing professional* (pp. 1–13). Sense Publishers.

Leung, S. S. (2012). Towards a model for parental involvement in enhancing children's mathematics learning. *Proceedings of the 36th conference of the international group for the psychology of mathematics education*, Vol. 3–107. Taipei, Taiwan, 2012/7/18–22.

Leung, S. S. (2013). Teachers implementing mathematical problem posing in the classroom: Challenges and strategies. *Educational Studies in Mathematics, 83*(1), 103–116.

Leung, S. S. (2015). *Math after dinner: enjoy family math time.* Outside The Box (OSB) Publishing Co., Ltd. (In Chinese; 140 pages).

Leung, S. S. (2018). *Parental involvement in children's mathematical learning: Why and how parents attend to family math tasks?* International Conference on Education and Learning (ICEL). Tokyo, 22–24.

Leung, S. S. (2019). *Hi! I am a fun math game.* Hunan Scientific Technology Publisher (in simplified Chinese; 132 pages).

Leung, S. S. (2023). Parental involvement in young children's mathematical learning: The story of family math tasks. In J. Cai, G. J. Stylianides; P. A. Kenney (Eds.), *Research studies on learning and teaching of mathematics.* (pp.157–178). Springer.

Leung, S. S., & Lo, J. J. (2010). *Sweet play.* Teaching Children Mathematics. NCTM. 16 (6), 330–331.

Leung, S. S., & Wu, R. (2000). Sharing problem posing at home through diary writing: Case of grade one. *The Australian Mathematics Primary Classroom, 5*(1), 28–32.

Maher, M. (2007). Home-school partnership within mathematics intervention. *Australian Journal of Early Childhood, 32*(3), 48–58.

Ministry of Education (MOE). (February 2014). *Principles and guidelines for curriculum development of twelve-year compulsory education].* http://www.arteducation.com.tw/upload/201407 11165703.pdf.

National Council of Teachers of Mathematics. (2000). *The Principles and Standards for School Mathematics*. Reston, VA: Author.

Stevenson, H. W., & Lee, S. (1990). *Contexts of achievement: A study of American, Chinese, and Japanese children*. University of Chicago Press.

Van den Heuvel-Panhuizen, M., & Buys, K. (2008). *Young children learn measurement and geometry*. Sense.

Van den Heuvel-Panhuizen, M., van den Boogaard, S., & Doig, B. (2009). Picture books stimulate the learning of mathematics. *Australian Journal of Early Childhood, 34*(3), 30–39.

Yin, R. K. (1994). *Case study research design and methods: applied social research and methods series* (2nd ed.). Sage Publications Inc.

Chapter 13
Parents' Involvement in Math Problem Posing for Math Trail and Fun Fair: Examples From Grade Three and Grade Four

1 Introduction

As in the preceding chapter, this is also part of a larger study when a mathematics teacher educator (MTE, named Mytle) worked closely with elementary school teachers and conducted a case study on parental involvement in an elementary school near the university where she taught. The project extended its inclusion from kindergarten to primary six, to empower parents so that they were resources for children's math learning. There were multiple attempts over the years: Family math camps, parents' book reading study groups, math trails, fun fairs, and, picture books for young children. A detailed description of the funded project for eight grade levels (K1, K2, grade 1 to grade 6), its setting, and the development of family math tasks (FMT) was given in Chapter 12 of the book and in Leung (2023).

The MTE, Mytle, believed that to make children learn mathematics well, work with teachers was insufficient. She must also work with parents (Leung, 2018). An example of her involving parents is written work like writing math diaries (Leung & Wu, 2000, Chapter 11 of this book). Besides the completion of written work and arranged for parents to play math games in regular math camps (see Chapter 12 of this book), parents can be involved in outdoor activities.

When the activity is outdoor, there is a need to recruit more adults and it is best to include parents. Safety can be one reason but Mytle added two more reasons. First, when parents join the school's activity then they know more about the learning of the children. Second, when parents join, children will be excited that parents are accompanying them. Mytle believed that in order to empower parents, she must recruit parents to share responsibilities in the preparation for children's learning activities. Mytle agreed that parents were resources, they were not just helpers with arranging library books or doing gardening (Civil & Bernier, 2006).

In this participating school, Mytle read that Fun Fair is a tradition and the date was marked clearly on the school calendar each year. However, the games were non-math. Knowing this tradition, she negotiated with the responsible teacher and

S. S. Leung, *Mathematics Problem Posing in Action*,
https://doi.org/10.1007/978-981-96-8480-9_13

proposed to add a math component to the regular school event. She shared the book (Leung, 2015) to parents as a reference for children's math ideas, to pose problems for (1) a Math trial, and, (2) a Fun Fair math stall. To delimit the presentation of parental involvement in problem posing, only the part on the math trail and fun fair for grade 3 and grade 4 will be reported here. The research questions were:

1. What types of math problems could parents pose in a math trail? What lessons did parents learn from the multiplicity of responses from all children, completed with the company of parents?
2. What lessons did parents learn from participation and division of jobs in a fun fair?

2 Background Literature for Actions

2.1 Involving Parents in Children's Math Learning

Parental involvement refers to participation from a parent when it comes to schooling and her child's life. In the US, the Center on School, Family, and Community Partnerships, at Johns Hopkins University works towards PI and divides school parent involvement programs into six broad categories: (1) Parenting, (2) Communicating, (3) Volunteering; (4) Learning at home, (5) Decision-making, and (6) Community collaboration (Epstein, 1995). Among the six, learning at home (Category 4) ranked highest; while the others appeared to have less direct effect on student achievement (Epstein, 1995). In South Taiwan, learning at home also ranked highest among the six when 958 parents with grade one children completed questionnaires and, when the items were geared toward mathematics subject (Leung & Chou, 2012). A variety of ideas for involving parents in math learning at home were given in a collection of articles published in two official journals of NCTM (see Edge, 2000).

What kind of math learning can happen at home? Contextual activities at home can enhance mathematics learning (Boaler, 2009; Perry et al., 2007). For example, watching counters in a microwave oven is a good way to learn backward counting on numbers; pairing chopsticks in the kitchen is a way to learn counting by "2"s; and, recording the number of pets seen for a car ride from home to school during weekdays to learn tallying. Therefore, the reason for the few math-related activities for parents is not that the activities are not available. Parents might miss the opportunities or do not know they are math-related. Even when the home was provided with both math and non-math-related activities, young children tended to be engaged in non-math-related ones (Tudge & Doucet, 2004). Moreover, even when math-related activities were recorded at home, they were solely on counting and very few extensions to activities such as addition and subtraction or further (Tudge & Doucet, 2004).

What are some explanations for not having math or not having math other than just counting and simple ideas at home? Could the young children or their parents be the reason? Supervising young children's homework often creates tension between children and parents due to the absence of know-how in parents (Leung et al., 2012). If a parent reverts to attending to children's math learning what possible activities can be arranged for an adult and a child? For example, parents can volunteered to play math games in their son's school while teachers were having staff meetings and observed how their children and peers learned through games (Leung, 2008).

Apart from written work at home or indoor board games what else? It is left to study if parents and children can be engaged in outdoor activities instead.

2.2 Connecting Math Ideas in Outdoor Activities

In school, most of the learning and teaching take place indoors: inside the classrooms, going to laboratories, cooking in a kitchen, and staying indoor constitutes most of the instructional time for the students. However, outdoor activities allow for more exploration. For example, children may walk out to the school garden to watch the plants. In math, what kind of outdoor activities can children engage in, to learn, consolidate or explore new ideas? For example, students doing "math walks" reported that the activity provided a powerful opportunity for informal mathematics learning (Wang & Walkington, 2023).

In Taiwan elementary schools, two common outdoor activities are walking trails and fun fairs.

A walking trail was first conceived as an outdoor activity to help students become active learners by exploring ideas and concepts that exist in their communities. Walking is typical. Unlike stalls in the fun fair that were next to each other, stations on a walking trail are far apart and children/adults must read a map and walk there. A worksheet can guide the participant to complete a task or solve a problem once they walk to that particular station. In each station, there is a station manager who checks the completion of the tasks. The task usually requires the player to use a landmark or feature of the land. Therefore, the task cannot be done elsewhere.

A school fun fair is an event organized every year where students, teachers, and parents come together to have a great time. The fun fair usually includes various games, food stalls, rides, and activities to entertain participants. Usually, there are different stalls and one stall can convey a specific theme. Children and adults can visit any stall they feel free to and have some fun from the visit. If the stall is for math, then the design will be a completion of a task, and upon completion the player earns some scores (in a passport) or prizes. If the difficulty level is not fixed, then anyone can visit the stall. It is customary that the tasks are simple or at least not too challenging so that visitors can leave the stall with pride and prizes. The school will have donations to run this fun fair. In all, the aim is to make all participants happy.

Math is all around. Math ideas can be used to develop tasks for math trails and fun fairs. According to the curriculum guidelines for mathematics (MOE, 2018), children come across multiple topics in different strands such as numbers, measurements, shapes, and space As children study up to grade 3 and grade 4, there is the part on numbers that includes whole numbers and computations in four algorithms; and, the introduction of fractions and decimals. For measures, they already learned about money, length, time, angle, and also perimeter/area. There is also an introduction on capacity. For Shapes and space, there come perimeter and area, estimations, right angles, rectangles, and squares. Children need to be provided with a variety of experiences to master such concepts. How about walking outdoors to visit buildings and trees in the playground to explore shapes and space, length, area, and perimeter?

The above suggests a background for involving parents in children's math learning from home to school and from indoors to outdoors. According to the curriculum guidelines (MOE, 2014), there are three core competency dimensions for all subject areas. They are: (1) Autonomous Action; (2) Communication and Interaction; and, (3) Social Participation. Under these three dimensions for math subjects, suggestions for elementary grade levels are listed. Examples are: Students formulating plans and translating mathematical solutions into everyday applications, perceiving mathematical forms or patterns in works of art, and, working with others and respecting different ideas to solve problems. Can children learn these math topics and attain core competency dimensions? Can these be accomplished outdoors? What and how can parents contribute?

3 Method

3.1 Participants and Setting

The activity, a family math camp, was held in an elementary school near the math educator's institution year round and for all grade levels (K-6). The participants met every other Friday. Participants included teaching team members ("teachers") and family math campers ("learners" were parents and children). Activities were completed in classrooms and were mostly board games. The team included Mytle (a math teacher educator), three elementary school teachers, and three teaching assistants. For grade three and grade four, parents' attendance record was about 50. Parents, mostly mothers, were from a diverse background. Most of the parents took a leave from work to attend this camp. No child could attend without at least one parent or a guardian present. Tasks called family math tasks (FMT) were specially designed so that it could not be completed without a parent (see Leung, 2023).

For the design of the Math Trail, Mytle invited parents to walk around the playground, made decisions on features in the school, drew maps, and finally, parents posed problems for children to walk through the trail and complete. For the design of the Fun Fair, parents worked on the ideas during the Friday meetings, and shared the work. Mytle and the research team did minimal input.

3.2 *Materials for Math Problem Posing*

At the beginning of the school year and for each term, the teacher educator sent, through the class teacher, a letter to invite parents to attend the family math camps (met every other Friday). Each time the module was different but matched the academic planning for the teaching schedule in math. Since there were four schools, the teacher educator studied all textbooks used by the schools and she made sure that there were two math ideas commonly covered by all four schools over the 14 days. The contents included in the module (Grade 3, Grade 4): Multiplication and division, Lengths, Capacity, Fractions and Decimals, Estimation, Big numbers, Perpendicular and parallel, Area and perimeter, Time, and, Unknown. As above, the list of activities and events given in the invitation letter matched what the math contents taught over the year. Among the camps, students and parents did problem solving and problem posing. However, the focus of this chapter is on how parents involved by doing problem posing for their children. At the end of the first term, it was a math trail, and, at the end of the second term, it was a fun fair. The two marked a happy ending for both children and parents. Certificates were also given.

3.3 *Data Collection and Analysis*

For parents, the math educator conducted focus group interviews with parents during the first session. Parents also returned diaries and teacher educators kept journals. For students, mainly written work, in the form of worksheets or diaries. Students were divided into five groups (A to E) and completed in the order assigned but for fun fair children were free to complete the four tasks in any order they like.

3.3.1 Sunny Math Trail Worksheet

<u>Sunny Math Trail</u>

1. Boys and girls please complete this math trail happily, with your dad (mom)
2. See which group can complete this super mission
3. When you answer, you can use your hand or textbooks as the answer units, not necessarily in units like cm
4. Note: when complete, go to "Wall of Home" for a gift and wait to take a group photo

Your route.

Group (A) Please walk in this order: 1st, 2nd station → 3rd, 4th station → 5th, 6th station → 7th, 8th station → 9th, 10th station (group B, C, D, E start from 3rd, 4th station and so on).

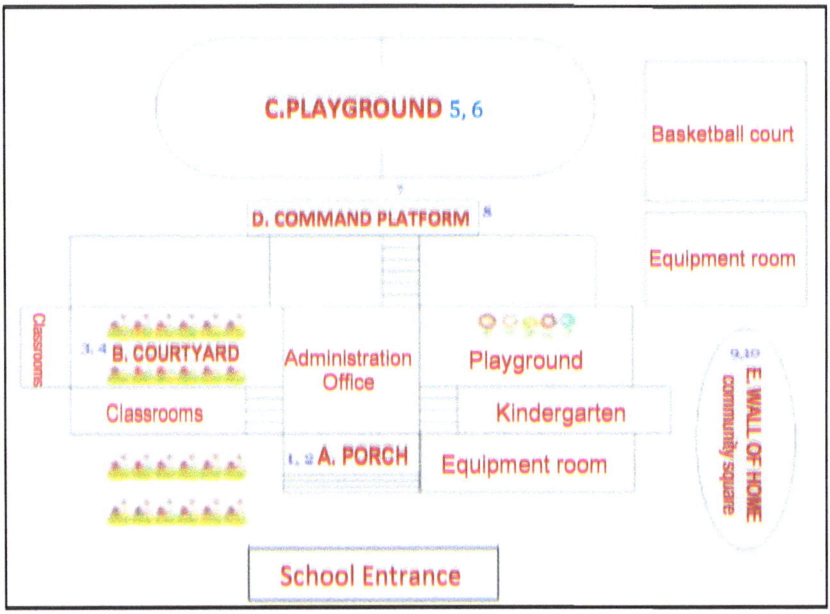

Special thanks go to our dearest parents. They spent time posing and revising until the final version was made.

*Dear boys and girls: After you complete, write what you feel inside this box below. You may also draw what you think about today's math trail. Mytle ^.^*

3.3.2 Jolly Fun Fair Passport

There are four stations and the passport was used for children to collect a "stamp" when they complete one task. There was no score for each task, just got one check ("V") upon completion of one task (up to four).

4 Results

As in the other grade levels, Mytle offered parents workshops for enhancing learning math through interactions between/among family members. However, in addition to board games, outdoor activities were added for grade three and grade four. Parents initiated efforts from math content to activities, and, also prize giving. Mytle purposely recruited prospective teachers to help and thus learned how to work with parents and children. In this chapter, the author delimits the report to only include what and how parents pose problems and participate in Sunny Math Trail and Jolly Fun Fair.

4.1 The Sunny Math Trail

4.1.1 Parental Involvement in Math Problem Posing

The team members went for a walk in the school playground while the children went for math games with prospective teachers. Parents took photographs of potential features on the campus and returned to the classroom to pose math problems. In choosing stations for the Math Trail, there were several concerns. The first concern was about the *weather*. About the stations, parents' "*not too sunny*" was a concern as "*Not too warm there*". The second concern was safety. They preferred a long safe route instead of a short one yet "*might be dangerous*". The third concern was spreading out players on the campus. The teacher educator reminded them that the stations should not be close to each other so that children would be evenly spaced out during the event. "*Yes, not making them bump against each other...*" After using these concerns for considering the stations, there were five locations they decided to use. Parents were then divided into 5 groups (A to E) and each group was brought to a final of two stations with four problems. Their ideas were passed around (A- > B; B- > C, C- > D, D- > E, E- > A). They did an oral presentation during the meeting.

Math problems were initially posed and decided until after a process of *inclusion, exclusion,* and, *revision. Inclusion*: data showing parents' preference. The top criteria were interesting, the second was not so hard yet needed some thought, and the third came to the relatedness to math levels (covered by all four schools). However, one interesting idea came during one meeting. "*How nice, there are multiple ways to solve this problem.*" The fourth criterion was having multiple ways to solve the problem.

Exclusion: Parents made choices by elimination too. When the two were alike, they picked a more attractive context for children. When problems were equally good they flipped a coin. Parents also excluded problems that could be done by just reading the worksheet and no need to walk to the location. *Revision*: Mainly on wording, "*My son must misunderstand if we do not change this line…*" They revised problems and difficulties and also took advice from Mytle. After whole class discussions, a set of tentative problems resulted. The teacher educator revised it further and passed it to class teachers for reading and checking. Mytle and the team used results on parents' MPP to make worksheets.

4.1.2 Problems Posed by Parents and Solutions Given by Children

There were ten stations. The presentation of results in each station includes a box with a photo of the station and the math problems posed for players to complete when they walk there. Parents reviewed and considered them as interesting, not too hard nor easy, and, allowed multiple solutions. Parents remembered that their children already learned length, capacity, area, perimeter, estimation of quantity, and geometric shapes. Mytle checked if the problems required math that children had not learned before, and also adjusted the problems so that the time spent on all five stations was about the same. "The time use is important. Let us make sure children would move to the next station at the same time."

On that day, parents accompanied the children in completing the walking trail. The weather was great and the timing was correct.

After the event, returned worksheets were read over the meetings. Parents were excited to read a multiplicity of responses from all children instead of just the answers given by one particular child. "*It was great that we posed problems that allowed multiple ways to solve…*" Underneath each box were children's solutions and analyses by parents and Mytle. The team also gave positive feedback to the children.

A. PORCH (1, 2)

	1. Please observe clearly and answer the questions (1) Please measure the red circular chair using your way. What is the perimeter? (Jenny's mom) (2) How many people can this chair accommodate? (Benny's mom)

(continued)

(continued)

	2. This is a bench (1) The face of this bench has __ sides.How about the perimeter? It is__cm long. (Kim's mom) (2) Ms. Leung wants to put this bench into a large box. How large is the box? How wide is the box? How tall is the box? (Ming's mom)

Question 1 asked for the size of the circular chair and the answer can be any units the child and parent can think of. They normally used the length of palm, pencils, or the length of a side of the worksheet. Some pairs bought rulers and used standard units like cm. They suggested 10 children sitting on this round stool. Question 2. The bench is an octagon, all answered correctly. The suggested length, breadth, and width of the box could be any dimensions greater than the actual size of the bench, as the question did not say "minimum" size.

B. COURTYARD (3, 4)

	3. Please observe clearly and answer the question (1) How long is the iron handrail? You can calculate using your way. (Kate's mom) (2) Look at the picture. This is made up of two geometric shapes. They are__and___. (Sun's mom)
	4. Please observe clearly and answer the question (1) How tall is the iron fence? You can calculate using your way. (Ling's mom) (2) Watch the region encircled by this red dotted line. How many squares? Think carefully. A big square is also a square, not just a small one. (Kay's mom)

Question 3 asked for length and shape. They completed this task by using objects at hand. One child used a writing pad (*33 times the length of my writing pad*) and their methods were admirable. "*I say 24 times the length of my foot.*" When they named the geometric shapes, they might use the names learned in math lessons. Whether or not they came across the names in school (*sector, trapezium*) they somehow managed to name them using "*fan shape*" or "*library stand shape*". Question 4 asked for length too. They wrote *7 handspans, 84 cm, and 6 pencils*. One student wrote an approximation, "from 80 to 100 cm" and his mom sent feedback, "*That's correct! ^0^*". With the hint in Question 4-(2), all counted carefully the number of squares and were correct. "*8 squares*". Some explained "*6 small and 2 big, altogether 8 squares.*" Parents replied. "*Thanks for showing us how.*"

C. PLAYGROUND (5, 6)

	5. Please use your way to answer the questions (1) If I walk in the playground from H to K, how many footsteps for adults? How many foot-steps for children? (Cloud's mom) (2) What is the width of the playground's runway? (Chuck's dad)
	6. Please observe the picture clearly and answer the questions (1) How many right angles are there? (Truth's mom) (2) There are __squares, __rectangles and __triangles. (Tommy's mom)

Question 5 asked for the length of one walk from H to K of the runway. The parents who posed this problem was a daddy who expected children to walk with parents and answer using how many steps. However, each parent–child pair stood on different parts of the runway (using the map with marks H, and K) so the answers were quite different, depending on where they thought the mark H and/or K were in the real playground. "*As long as the attempts were there, these answers were all considered correct.*" The second part was answered by using the length of the string of the name tag. Children used convenient objects they had at hand. Question 6 first asked the number of right angles and they replied, "34". Question 6-(2) asked for the names of the shapes in the skip/jump ground. A child might or might not count squares and rectangles separately (all squares or all rectangles). There were mistakes in this part and parents wrote, "*Please try again.*" instead of saying that the answer was wrong, and wrote the correct answer next to the feedback.

D. COMMAND PLATFORM (7, 8)

	7. Please look at the picture and answer the following questions. What shape is it?_________ (Rainbow's mom) (1) How many identical shapes make up the wall of this building? (2) If we group in threes, how many groups can we make?_________ (Fung's dad)
	8. Please look at the picture and answer the following questions (1) What is the area encircled by a red dotted line? Please use your way to measure and answer. (Soul's mom) (2) What geometric figure is encircled by this red dotted line? (Vivian's mom) (Children in second grade and third grade can discuss with your parents!)

Question 7 is the side of a building with three rows of four rectangles. However, one-third of the children thought they were squares. Watching the real building and reading its photo might cause the difference. Mytle remarked, "*That all squares are rectangles and how to differentiate was a bit hard for children at this stage*" and reassured the parents that there was no need to worry." Question 8-(1) asked for the area of the region shaded to be a dotted line. There were several methods: adding up or finding the outer region and then deducting some parts not expected to be included. Soul's mom was excited to read the children's giving a variety of answers: Counting the number of bricks, using writing pads, estimating as one square meter, and, five steps times five steps. Question 8-(2) asked for the shape and parents read mostly "*square*" and "*rectangle*". To Vivian's mom's surprise, she reported two worksheets with, "*rhombus*". The most wonderful feedback the parents agreed on was, "*Dear children, next time you try to imagine reading the shape as a bird flying above!*".

E. WALL OF HOME (9, 10)

	9. Please observe the round arch (1) Please use the dwarf tree beside the wall to estimate how many dwarf trees will make up the wall. (Chin's mom) ______________ (2) How tall is the dwarf tree? How tall is the wall? (Sally's dad)
	10. This is a stylish bench (1) How many strips of wood make up this bench? ______ How long is the tripe of wood? _________ (Ming's mom) (2) If people can sit on both sides of this bench, how many children can sit on the bench? (Kimberly's mom) ______________

Question 9 was about estimation. Their methods included using the height of a tree or the height of the child instead. "Eye estimate" was emphasized and children closed one eye when they completed this part. Question 10 asked for the estimation of the length of a piece of wood, and, to tell how many children can sit on both sides. Again they used handy objects to complete and were creative. "*Tree is about 50 cm tall*" "*Wall is 5 times as tall.*" Parents wrote, "*That was smart to use about*".

II. Expression of feelings after finding the answers.

Other than answering the worksheets they wrote an account of what they thought about the walk, walking through the math trail. Children's feelings were (1) about math, (2) interaction with family members, and (3) expressing feelings or saying thanks explicitly in words. The worksheet design purposely suggested children to thank the parents. Children considered walking through a math trail as very interesting, and fun, and hope to walk on more math trails in the future. Finally, they expressed their gratitude to their parents. Then also thanked Mytle (the math teacher educator) and the teachers. Some excerpts are given below.

MATH

"The shapes on buildings are not the same when viewed from the side." "My estimation was better when I used a piece of string and not my fingers." "The word perpendicular is tough, it is referring to right angles."

INTERACTION WITH PARENTS

"It was fun reading Dad's number of steps was not the same as my number of steps, although the length we measured was the same. Then I realized that his step is so much longer and therefore less number of steps.

FEELINGS OR THANK YOU

It is funny! I will join next time! I liked outdoor activities. Mytle gave us a very different math class. It was delightful!

"Thank you, Mom and Ms. Mytle, organizing this math camp must be tiring.

 ***I am happy because we have completed***	 ***It was fun to walk together with Mom***	 ***Thank you, teachers. You all worked so hard.***

"Thank you: Mytle and other teachers, you all helped me to learn more information about the school campus and math."

^0^ I love Ms. Myle!"

Parents expressed that they were happy to walk and find answers (although they knew the answers). They were overjoyed to see their children being very eager to find answers and helped each other so that all the friends in the class finished in time. "*Ms. Mytle said we cannot take the group photo until all groups arrive at the endpoints.*"

4.2 Jolly-Math Fun Fair

4.2.1 Problems for Jolly-Math Fun Fair

First, parents and teacher educators gathered for a discussion on the development of materials. Then Mytle read the posed problems for the fun fair and the team selected four problems. Then they were divided into four groups. Each group worked on one problem. They tried the problem as if they were children, and then, they revised it according to the correct/incorrect solutions. When all were satisfied with the problem they changed to another group. The process was repeated until all parents agreed on all the problems. The final version of the set of four problems was made.

Second, parents met to talk about their concern for the administration. Although the work was usually done by a responsible full-time research assistant of Mytle's project, parents expressed that the work be divided into jobs carefully and they also shared responsibilities. They were considering which parents to take turns to be fun fair managers of the four stalls so that they would not be bored. The work included designing posters, printing passports, wrapping prizes, decorating stalls, and tools used for each stall. Finally, on the big day, children registered got passports, and then planned their route. They visited the stalls individually or with friends freely. As the tasks were not difficult they all completed and were rewarded a prize donated by eager parents.

1. Into a Magic Maze (3 rows by 3 columns) Children needed to fill in 9 numbers. When adding the three numbers horizontally, vertically, or slanted way, the 3 numbers are added to the same total. Note: the numbers were two-digit numbers	
2. Go for Crazy Fishing One player threw the dice to determine the operation (+- × ÷), and the other picked up the fishing rod and started fishing within 5 s. The two arranged fish (with numbers on it) on the table, and used the operation cards to add, subtract, multiply, or divide. The winner was the team who made the greatest number	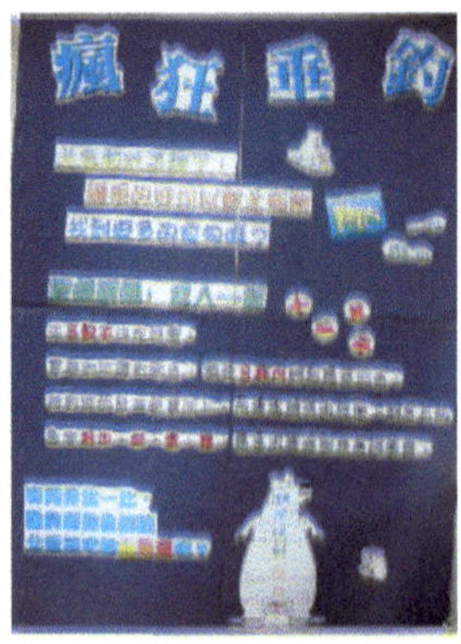

(continued)

(continued)

3. Fun with Cups of Wonder Two cups of different sizes (2 units, 3 units) were given. Children were invited to make up different amounts of water using the two cups. For example, they use "2 + 3" to make "5". Later, they were given three cups (2, 3, and 5 units) and were told to try making up different totals. For example, they use "2 + 3 + 5" to make "10". A tank of water was there for them to use for checking	
4. Go for Star Mystery Throw the dice into the big box, and the number thrown is the password of the planet (1–36). Draw a planet card with the numbers "1" to "9" written on a three-by-three grid card, and use + - × ÷ to calculate the password (e.g. "15") by using the three numbers in the vertical row, horizontal row, or diagonal row	

4.2.2 Lesson Learned and Parents' Feedback

1. When playing games, there is no difference in age

Both adults and children love games and their age appeared to be the same, especially when they played as opponents. "*Today, we are also ten years old, you see…*".

2. Minus means more: sharing ideas with others means resulting in more ideas.

"*When you think you subtract ideas from your brain, you do not lose ideas as more ideas are added to your brain.*"

3. The art of revision

Parents found peer review extra helpful. "*Revise, revise, revise. After peer review and revisions, you can change nothing to everything.*"

4. The Golden principle: working with hands and the brain

Parents found that for math games that involve children using their hands, children also use their brains.

5. Action

Finally, they found that taking action is better than just proposing. "*Let us just try it out right now. The idea should not be frozen*".

Mytle and the team attended the two staff meetings of the school after the two events. Results of the above two outdoor events suggested that the participation of parents made a great difference. What constituted the main part of parental involvement in posing problems for two school events? The stamina was the friendship developed over the past workshops (family math camps) of the term. They worked and helped together on family math tasks (Leung, 2023) and remembered the successes and failures of solutions from their children and themselves. Parents also considered themselves learning a lot from reading math textbooks, to posing math problems, and, trying to imagine the different ways to solve the problems. "*It was fun reading how children solved those problems, especially in ways that we did not expect when we posed and solved them.*" For the fun fair, since most of the tasks were completed without any mistakes, and, nearly all children finished and got prizes. "*The sense of achievement was great and it was lovely to hear them saying thank you when they returned home.*"

5 Conclusion

Research supports claims about the importance of home experiences in children's acquisition of mathematics (e.g. LeFevre et al, 2009). In addition, outdoor activities such as math trails were also beneficial to children children's learning, including their personal, social, and emotional development (English et al, 2010). Parental involvement in children's math learning was re-visited here and in this case, MPP was used in outdoor activities. In Leung (2023), the program for parents was laid out and examples of young children's family tasks were given. In this chapter, parents were doing problem posing with target problem solvers—their children. They were teachers posing problems and considered contents and levels of difficulty. When in need of help, MTE shared what she knew about the three goals in the national curriculum guidelines (MOE, 2014). Learning math (and the other four domains) should be geared towards (1) Autonomous Action; (2) Communication and Interaction; and, (3) Social Participation. The findings in this study showed that engaging parents (adults) and children in activities like Math Trail and Fun Fair math stalls enabled participants to achieve the above three goals. Parents reading worksheets of children were able to know examples of children formulating plans and translating mathematical solutions into everyday applications. The investigator also concluded that combined efforts enabled children to attain goals in the math curriculum standards such as perceiving mathematical forms or patterns in works of art, working with others, and respecting different ideas to solve problems. Parents were aware of the materials children learned over the years (number, measurements, space) and possible matches to objects (Staircase, jump plane) in the school. It was found that organizing the Fun Fair and Math Trail provided invaluable learning opportunities to all involved participants. All of them were co-learners (Jarworski, 2008).

References

Boaler, J. (2009). *What's math got to do with it: How parents and teachers can help children learn to love their least favorite subject*. Penguin Books.

Civil, M., & Bernier, E. (2006). Exploring images of parental participation in mathematics education: Challenges and possibilities. *Mathematical Thinking and Learning, 8*(3), 309–330.

Edge, D. (2000). *Involving families in school mathematics: Readings from "Teaching Children Mathematics," "Mathematics Teaching in the Middle School," and "Arithmetic*. Reston, Virginia: National Council of teachers of mathematics.

English, L. D., Humble, S., & Burnes, V. (2010), Trailblazers. *Teaching Children Mathematics, 16*(7), 402–409.

Epstein, J. L. (1995). School/family/community partnerships. *Phi delta kappan, 76*(9), 701.

Jaworski, B. (2008). Mathematics teacher educator learning and development. In B. Jaworski & T. Wood (Eds.), *The mathematics teacher educator as a developing professional* (pp. 1–13). Sense Publishers.

LeFevre, J., Skwarchuk, S. L., Smith-Chant, B. L., Fast, L., Kamawar, D., & Bisanz, J. (2009). Home numeracy experiences and children's math performance in the early school years. *Canadian Journal of Behavioral Science, 41*(2), 55–66.

Leung, S. S. (2008). *Math wonder bag: Get out! Math sleepy bug*. Kaohsiung: Outside The Box (OSB) Publishing Co., Ltd. (In Chinese, 174 pages).

Leung, S. S., & Chou, H. C. (2012). A survey study on parental involvement in mathematics learning for elementary school children. *Proceedings of the 36th Conference of the International Group for the Psychology of Mathematics Education, Vol. 4–294. Taipei, Taiwan: Author.*

Leung, S. S. (2015). *Math after dinner: Enjoy family math time*. Outside The Box (OSB) Publishing Co., Ltd. (In Chinese, 140 pages).

Leung, S. S. (2018). Parental involvement in young children's mathematical learning: Why and how parents attend to family math tasks? *2018 International Conference on Education and Learning (ICEL)*. Tokyo, 22–24.

Leung, S. S. (2023). Parental involvement in young children's mathematical learning: The story of family math tasks. In J. Cai, G. J. Stylianides, P. A., & Kenney (Eds.), *Research studies on learning and teaching of mathematics* (pp.157–178). Springer.

Leung, S. S., & Wu, R. (2000). Sharing problem posing at home through diary writing: Case of grade one. *The Australian Mathematics Primary Classroom, 5*(1), 28–32.

Leung, S. S.,Wu, P. J., & Chen, C. H. (2012). A study on parents attending math study groups was designed by a grade one elementary school class teacher. *Proceedings of the 36th Conference of the International Group for the Psychology of Mathematics Education*, Vol. 4–293. Author.

Ministry of Education. (2014). *Curriculum Guidelines of 12-Year Basic Education General Guidelines* (in Chinese)

Ministry of Education. (2018). *Curriculum standards for 12-year compulsory education-mathematics*. Ministry of Education.

Perry, B., Dockett, S., & Harley, E. (2007). Learning stories and children's powerful mathematics. *Early Childhood Research & Practice, 9*(2). Retrieved from http://ecrp.uiuc.edu/v9n2/perry.html

Tudge, J. R. H., & Doucet, F. (2004). Early mathematical experiences: Observing young Black and White children's everyday activities. *Early Childhood Research Quarterly, 19*(1), 21–39. https://doi.org/10.1016/j.ecresq.2004.01.007

Wang, M., & Walkington, C. (2023). Investigating problem posing during math walks in informal learning spaces. *Frontiers in Psychology, 14*, 1106676.

Chapter 14
Parental Involvement in Reading Picture Books for Promoting Mathematical Thinking and Problem Posing: Two Examples on Young Children

1 Introduction

In a competitive atmosphere for countries nationwide, one must emphasize educating the next generation for the future. A question that is frequently asked yet remains unanswered: How can children study mathematics without pressure yet keep up with the standard from a country's curriculum standards and as benchmarked on literacy in large-scale studies like TIMSS (IEA) or PISA (OECD)? In Taiwan, mathematics is a subject domain that is to be studied until grade 12. However, it is often deemed to be difficult and its learning requires help from teachers and also parents. How can parents get involved in children's math learning? Parents did not learn math the same way as children do nowadays. A way is to participate in parent workshops, where the investigators developed research-based family math tasks (as in Leung, 2023), and evaluated the potential for enhancing parents' involvement (e.g., attending math camps together with children, and trying tasks at home). With this objective, a study on families with young children was launched by Mytle, a math teacher educator, as part of a larger project (from K to grade 6).

According to PISA (Program for International Student Assessment), mathematical literacy is concerned with "the capacity of students to analyze, reason, and communicate effectively as they pose, solve and interpret mathematical problems in a variety of situations involving quantitative, spatial, probabilistic or other mathematical concepts" (OECD, 2004, p. 37). The emphasis on math thinking, problem posing, and problem solving has always been emphasized in curricular documents (NCTM, 2000; MOE, 2003, MOE, 2018); including the part on having children do mathematical thinking and communication. In problem posing, there were opportunities having students to think about a situation, mathematize, and pose problems (Silver, 1994). However, engaging young children in problem posing is not easy. One suggestion is to use rich contexts in picture books because young children love stories and often add emotions and imagination to reading (Griffith & Clyne, 1991).

S. S. Leung, *Mathematics Problem Posing in Action*,
https://doi.org/10.1007/978-981-96-8480-9_14

Hence, combining problem posing and picture books is a promising avenue to investigate math thinking. However, studies from the psychology of mathematics education community focusing on K to grade 3 in the three decades (from 1995 to 2005) only made up less than 10% of reports in PME (Psychology of Mathematics Education) meetings; and limited attention was paid to cross-cultural studies on young children (Mulligan & Vergnaud, 2006). Hence, the attention to more studies on young children's math learning is needed. Until recently, a systematic review (of studies from 2000–2022) attended to the use of picture books in math instruction (Zhang et al., 2023) and reported approaches, effects, and factors for effects.

Mytle believed that it is worth investigating how mathematical problem posing can be done through reading picture books. The aim of this study is to use this approach to investigate how parents read picture books together with young children, and how parents and children interpret situations together from picture books and pose math problems. We address a few questions: What happens when parents read picture books together with children? What math thinking can be inferred? What math problems will children consider?

2 Background Literature for Actions

Posing problems and writing math diaries are strategies for parental involvement in Taiwanese young children, to be completed in schools or at home (Leung & Wu, 2000). For young children, their early number learning is important and may predict future success in mathematics. However, whether they have any problem posing at home depends on the home environment and the interactions with family members, especially the parents. Parents need to find appropriate materials and tasks. Learning number concepts from reading picture books is a good start.

2.1 Children's Early Number Concepts and Development

Baroody and Purpura (2017) provided a foundation chapter for understanding young children's early number concepts or numeracy. Children's numeracy when they start learning in school predicts the rate of acquiring further mathematical knowledge (Jordan et al., 2007). However, the part on informal mathematical knowledge is also important before children go to school. There will be learning difficulties when there is a gap between formal instruction and informal knowledge or underdeveloped informal knowledge (Baroody, 1987; Ginsburg, 1977). Hence, strong early childhood mathematics education was needed to bridge this gap. Examples of effective early childhood math education were given, such as learning trajectories (Sarama & Clements, 2009 and big ideas (Baroody et al., 2006). Finally, the authors tabled out to explain verbal-based number sense, as a basis for developing what they called, "Fluency with basic sums and differences." The table included levels to augment textual explanation on stages in number development: (1) enumeration (E), which entails determining cardinal number; (2) numerical relations (R), which involves comparing two or more collections or numbers; (3) operations on number (O); and (4) verbal counting (C). Following is the part on basic enumeration, with verbal subitizing (Kaufman et al., 1949) and Cardinal numbers.

Among the stages the earliest one is enumeration, The principles in counting by Gelman and Gallistel (1978) are four: One-to-one correspondence; maintaining counting labels in a fixed order; order irrelevance, and, applying cardinality principle. These principles are universal among young children in different countries as they exhibited strategies like counting-all or counting-on, skip counting, and using fingers or number boards (Baroody, 1983). In Taiwan mathematics curriculum standards (MOE, 2018) the progression of number learning and development is from counting, cardinality, and ordinality, to multiples. This progression is about the same nationwide, although the contexts for young children's learning are culturally diverse.

2.2 *Mathematical Problem Posing and Picture books*

The art of mathematics making (Polya, 1945) and the habit of considering problems served as a backdrop for this study. Attention to research and practice on problem posing was warranted, and the effects of instructions were reported in previous PME meetings (Silver, 1994; Becker, 2003; Sáenz-Ludlow, 2006). Among the various tasks for children to pose problems, Sáenz-Ludlow (2006) used stories of their own experiences that resulted in creativity in children's numerical thinking, numerical relations, and solution strategies. Children's solutions surpassed the difficulty of textbook problems for their level. In another study, it was found that reading a book with an adult reinforced a sense of achievement, and adult praise could promote children's posing (English, 1997). These prior studies confirmed the potential of adults reading books in rich contexts with children skillfully. Having adults read picture books and then listen to children's comments or talk about the stories is a possible way to promote young children's problem posing. Recently, there was also evidence of the potential of picture books as an impetus in children's math learning (Hasemann, 2003; Van den Heuvel-Panhuizen & Van den Boogaard, 2008). With clear pictures and interesting stories, picture books successfully engage young children, both from the story itself and from the reading process.

The use of one book titled *Being Fifth* as an impetus for math thinking (Van den Heuvel-Panhuizen & Van den Boogaard, 2008), and a study followed (PALM project of Freudenthal Institute), compared the responses of young children in 6 countries). The investigators were looking at how reading with children and watching the utterances of children were related to mathematics. Children's responses had a high proportion that were math-relatedness. It was concluded that even the adult was not asking the child questions about math, and even though the book was not written for teaching math, this book proved to be an impetus for provoking children's math thinking. The study indicated that the choice of a great picture book was essential in encouraging children to think mathematically. And if without prompts or teaching, parents can easily use them at home.

2.3 Civil Framework for Analyzing Parents-Child Interaction

Research on positioning in mathematics education has attended both to moment-to-moment discursive positioning, often in the classrooms (e.g. Herbel-Eisenmann & Wagner, 2010); where discourse positioned students as followers of teachers' instructions. Studies with a positioning lens analyzed instances of positioning in classroom discourse and observed how learners are positioned to support learning. For example, when learners (mothers and children) are doing mathematics together the mother may teach the children right away or try to learn again the mathematics with the children (Civil & Quintos, 2023); positioning theory was used as a framework to analyse data taken from parent workshops, on how parents and children co-learned mathematics and on interactions between parents and children around mathematics. In the case of multiplication and perimeter, both mothers were learning together with their children. Mothers and daughters held the position of knowledge holders, and their positions changed in moment-to-moment interactions. The interactions showed instances of co-learning, that is, adapting and learning from each other. One daughter realized that the mom already knew the answer but tried to solve the problem as if she did not. The other, although having the answers different to her mom and a friend, tried so hard to position herself and explained what she did in school then finally expressed her thoughts so that she was accepted. It is worth studying when parents and children read picture books together and try to make sense out of the story, what math thinking and problems would be resulted?

3 Method

3.1 Settings and Participants

Participants were also parents from Mytle's funded research project on parental involvement in children's math learning projects (K-6). The parents attended meetings on Friday afternoons in an elementary school. There was a total of eight sessions a term and for two terms per year, where parents and children were engaging in a variety of math activities for number learning. After four weeks, the workshops made parents know each other and the children well. Parents who expressed an interest in storybooks were invited to be volunteers to tell stories for children in the school when teachers were gone for staff meetings. These parents became members of a follow-up study and their role was named, "*story-telling mom*". Among the story-telling moms, three were initially included in this picture book project (K, grade 1, grade 2). Adults were split to do one-to-one interviews when they read a picture book to an individual child. In this chapter, only two parent–child pairs were reported (K, grade 1). One of them was in kindergarten and one in grade one and were named Kate and Gia (pseudonyms). The story-telling-mom were Mika (Mother of Kate) and Meg (Mother of Gia). Since both of them were participants of the larger project and were familiar with each other, Mytle made up a time to meet the two parents together for a briefing. They were told about the aim of this study before they participated. A three-person group was made for text/voice contacts over mobile phones.

Note: The results, on grade 2, were reported elsewhere because the project was extended to include a set of four related picture books with more parents and children.

3.2 *Choice of Picture Books*

The first picture book is *Being Fifth* which was written by Ernst Junger. The book was used in this study for Kate and her mom (Mika). The second is *Counting Cats*, written by Marilyn Burns. It was used for Gia and her mom (Meg). These books are not intended to teach math but when the stories may provoke thinking of concepts like ordinal numbers, cardinal numbers, and multiples. The author clips below the two book covers and the stories, in brief, are given thereafter.

Being Fifth. The story is about a doctor's waiting room with five broken toys waiting outside. The title page shows that of the five toys, the wooden puppet is the fifth. When one repaired toy comes out, one goes in. The story is repeated until the fifth toy goes in. The last one to go in is the wooden puppet with the broken nose. Only then is it shown that the brightly lit room is a doctor's office.

Counting Cats. A girl keeps only one cat at home. When there are she gives away her cats because her mother allows her to keep only one cat. By the time she gives away her extra cats and only one left, a few cats may be born. She needs to check the number carefully. The ending allows readers to imagine what to follow. The story stops and it is left for readers to imagine what to follow.

The two books were new to these two young children.

Fig. 1 Book covers *being fifth* and *counting cats*

3.3 Data Collection and Data Analyses

Task-based interviews (Goldin, 2000) were administered. The two interviewers followed the guidelines about interviews with children thoroughly. Children were interviewed individually. Stories were read by an adult to each child and after reading, the adult and child talked about the stories. The adult reminded the child to re-state facts about the stories (Zobari & Gulley, 1989) and invited the child to predict (Van den Heuvel-Panhuizen & Van den Boogaard, 2008). When talking about ideas too far from stories, the adult repeated the child's response. On the next day, the adult talked with the child about the stories again and asked, "Guess what problem I am going to ask you using this story?" Children's guesses were audio-taped. If the child was quiet, the adult continued, "I also asked your friend to guess yesterday and they told me. What problems do you think your friend tell me?" The procedure was repeated and the child expressed their thoughts freely. Finally, the interviewers requested the child to record in a diary. The whole procedure (talk about story, posing problems) was completed in two consecutive days (Day 1: read and talk about stories; Day 2: recall stories and pose problems). There was no pre-set time limit, but none of the sessions went further than 30 min. In this study, the adult was the parent of the child.

Math ideas inferred from their recall and problems they posed were categorized by number development progression (Cardinality, Ordinality, multiples) and by sufficiency types according to Leung and Silver (1997). The five types of problems normally given by problem poser are: Not-a-problem, Non-Math, Implausible, Insufficient, and, Sufficient. Two children were interviewed as a pilot. After the pilot study, two parents used similar techniques and read stories with their children. In both cases (K, Grade one) the reading of each picture book (*Being the Fifth, Counting Cats*) was done at home.

Records of work from home were collected through carefully designed worksheets and to encourage sharing, the worksheets were named "*math diaries*". Parents were told that "*A diary can record an event or work of children after school, communicate, reflect, share results, emotions, queries, or make a thank you note.*" Mytle also explained the general principle of diary writing.

1. The adult talked about the take-home math activity with the child.
2. Parents and child wrote to share with a target reader in mind (e.g. Mytle, grandma).
3. The child may read aloud and the parent assisted in writing, or have the child write the reading aloud part using phonetic symbols.
4. Adult praised the child for making an effort using gestures, verbally, or in written forms.
5. Diaries were shared voluntarily but not compulsory.

Parent and child may express their feelings about playing math together at math using the communication boxes at the end of the worksheet.

4 Results and Discussions

4.1 What Happens When Parents and Children Read Books Together?

On parental involvement, Mytle noted a few strategies of the two participating parents. They (1) *found the right time to read and read leisurely*, (2) *checked what the children meant* when unsure, and, (3) *invited the child to participate* in the reading process. However, they were not aware of these strategies. Both mothers expressed that they did nothing extra and just read what was in the books. The strategies enabled two young children to talk about the books and freely express emotions and imagination about the stories.

(1) Parents and children *found the right time to read and read leisurely.*

Mika, "*Well, I do not think I have any strategy in reading. Usually, I wait for the right time when both of us were free, and just read about what's in the books. When we read, Kate is anticipating, always, she just loves to sit with me and listens like an angel.*"

Meg, "*Time is precious. We did not have a lot of time together after she went to grade one. I went to get her from school. Then I fed this hungry bird. She went out to play with her brother and also her dog. I read the book last Saturday. She was overjoyed whenever it was story time...*".

(2) Parents *checked what the children meant*

Kate exclaimed that the toys were sad. When asked for reasons, Kate replied, "*They are broken*". Kate said that she disliked the room. When Kate told Mika that she disliked the room Mika would not continue to read the next page. Instead, she would ask her why.

Mika, "*Why do you not like the room?*".

Kate, "*It is so dark.*"

Then Mika knew it was the darkness of the room.

Kate told Mika that she was not pleased with the frog. Mika asked, "*WHY?*" Kate did not use adult words like the frog was waiting impatiently, but Kate said, "*The frog sat upside down*".

Mytle responded at the three-person meeting, "*Checking what the child meant was a strategy, even though Mika said she did not have any strategy.*" Meg agreed, "*Instead of rushing through the pages to finish the story I would check if my child is following before I turn to the next page.*"

(3) Parents *invited the child to participate*

Gia was excited about the four newborn cats on the last page and asked her mom what their names were. The names were not given in the book. While not knowing the names and having doubts, she asked the child what the names could be...

Meg, "*Well, what do you think their names are?*"

Gia, "*You mean I better name the four cats.*" (excitedly)

Meg, "*Yes, please name them*" (relieved, happy)

Mytle responded at the three-person meeting,

"When not knowing how to reply, one strategy was to ask back and the child could answer what the mother could not."

4.2 Math Concepts Revealed After Reading Picture books (Day 1)

Parents reported what happened during the reading and talked about the story. They might opt to send written results in math diaries or verbally report the results to Mytle so that Mytle will record them in written forms.

Being Fifth. In this book, the chairs were lined up properly.

Sitting on each was a broken toy.

Mika: How many chairs?

Kate: 5 chairs.

She answered after counting, "1, 2, 3, 4, 5" and remarked that there were 5 in a row (i.e., a pattern of 5) when pictures showing a row of chairs. Kate occasionally stopped at a sentence or most of the time reading the picture and searched for key ideas. Mika witnessed all four principles of Kate in counting according to Gelman and Gallistel (1978).

First, the chairs and toys were arranged in a tidy row, Kate pointed accordingly and mapped one chair to one toy (one-to-one correspondence). Second, Kate counted in a fixed order, from left to right, and voiced, "1, 2, 3, 4, 5", maintaining counting labels in a fixed order.

Third, Mika reported that Kate knew that the order she counted was unimportant and the answer was the same (order irrelevance). Kate also saw a backward relation on the ordinal number (fifth to fourth...until being first) and made predictions successfully.

Kate, "*The first.*"

Kate, "*The third.*"

Kate, "*The fifth.*"

She could tell which toy was next, also, able to express using, "the first, the third, and the fifth." After one toy entered, she could tell the remaining number of toys waiting.

Fourth, Kate told Mika there were 5 toys (applying the cardinality principle). Finally, subitizing was possible for her. When Kate read the book cover she said there were five toys without counting.

Counting Cats. In this book, the cats were neither of the same size nor lining up in a row. In *Being the Fifth*, the preceding book, the chairs were of the same size and were lined up in a straight row and counting was easy. However, Gia heard her mom reading the book. When she read the page with cats inside the box, she said "5" with confidence. She could subitizing the total number of cats without counting. Gia reported on the number of cats quickly. When Meg asked her to check. She did it twice and counted in a different order. The answer was still "5". Thus, Gia demonstrated an ability to handle the Order-Irrelevance principle. Even when cats (on the same page) were counted in different directions she found that the number of cats was the same. Even when only part of a cat was seen behind another cat, she did not miss counting the cat.

As a grade one child, Gia correctly determined how many more. For example, by comparing "1 + 5" to "1" directly. Meg asked, "One cat, five more cats, how many cats more?" Gia knew "five more" without starting from "1" or adding '1' and "5' to get "6" then subtract "1" from "6" to get "5".

Meg, "*How many more?*"

Gia, "*Originally one, 5 came, so, 5 more cats.*"

Meg also witnessed Gia's knowledge of multiples and number reasoning, as Gia reported 20 eyes meant also 20 ears ("*I already counted 20 eyes; no need to count again, should be 20 ears.*"). Acknowledging that "*one ear to one eye*" was present. Gia also used adding–on strategies, When Meg changed the number of cats; child Gia reported 16/20 claws for 4/5) cats accordingly.

Meg, "*Let's count the claws*"

Gia: "*4*"

Meg, "*How about 2 cats?*"

Gia: "*8* (used fingers)"

Meg, "*What if there is one more cat?*"

Gia, "*Ok, 9, 10,* (not enough fingers, used Meg's fingers) *11, 12.*" "*12*"

In all, the two books successfully provoked math thinking in early number development, and the way the interviewer/parent talked through stories attracted children to think about or even think beyond the stories. When young children read books with rich contexts, evidence on subitizing, counting, cardinality, ordinality, and multiples can be inferred. Children's communications on ideas in the book were especially strong, as they predicted what would be happening next in the stories. They also remembered the stories very well, even on the next day. The ability to extend and remember the stories was witnessed by the interviewers/parents.

4.3 Problems Posed by Young Children in Problem Posing (Day 2)

As the children flipped the pages, they also predicted what happened and checked out their predictions. Upon turning the next page, Kate was excited to see that the toy had been repaired. For example, Gia remembered facts, such as the mother only allowed the kid to keep one cat. This memory of the story and prediction (anticipation) formed a basis for problem posing.

Children remembered the story from yesterday. She flipped the pages again.

Gia, "*If they keep '6' then there are '5' more.*"

On the next page came two more cats, adding to eight cats. Meg read...

Meg, "*If they keep '8'*"

Gia, "*Then there are 7 more.*"

This "read-predict" cycle was repeated. For example, when Mika and Kate read a toy entered the doctor's office. Mika asked Kate what happened after the toy entered the room and Kate predicted, "*Repaired*".

Being Fifth. Achievement in posing seemed to be influenced by the types of given information (Charalambos et al., 2003). As most young children had experience visiting the doctor, they asked reasonable questions or posed problems. Similarly, in this study, Kate asked "*How many chairs*?" or "*How many toys*?".

Mika, "*1, 2, 3, 4, 5, five chairs, and, let me count, five toys, am I right?*"

After Kate posed the problems, Mika pretended she did not know and "*worked hard*" in order to solve the problems.

Kate drew pictures of chairs to assist in posing a problem and, showed one–one correspondence (one toy - one chair), and understood cardinality-that there were five chairs altogether. Kate also asked questions (or posed problems) by analogy or by imitation, like, "*Who's next?*" When the fifth toy entered, "*Me In*", and no toy outside she still asked, "*Who's next?*"

Mika, "*Who's next?*" (Mika repeated Kate's problem since she still asked)

Kate and Mika, "*Ha ha ha*" (No answer)

Translation in English
Door open. One out. One in. Door closed. Now only four. Door open. One out. One in. Door closed. Now only three. Door open. One out. One in. Door closed. Now only two. Door open. One out. One in. Now only one. Then it is my turn. Door open. One out. Me in. Hello doctor.

(taken from the book, *Being the Fifth*)

Kate also asked questions relating to repairing such as how a doctor could help. As the answers were not given in the book. Mika and Kate told each other what they thought the doctor did. They exchanged ideas and accepted each other as long as the answers made sense. ("*Fix the toy*", or, "*Console the toy*").

By imitation or analogy, Kate asked "*What happened to the toys*" for Mika to answer. There were also non-math problems, "*Why is the lamp swaying to and fro*?" "*How long will it take*?" Interestingly, when reading this picture book, Kate asked, "How *many wings were missing in the penguin*?" or "*What happened to the wheels in the duck*?" and the observations were on the parts that needed to be repaired.

When the parent, Mika, asked, "*Do you like having a person who can repair them?*" the child, Kate, answered, "*Yes*" and then continued.

Kate, "*My dear mom too, she repairs everything for me.*"

Mika, "*That was a very sweet ending.*"

Apart from posing and solving problems using the stories, parents and children were enjoying family time after school.

Counting Cats. The most frequently asked question was, "*How many cats now*?" When Meg asked Gia to make a guess Gia was quiet. Meg remembered what Mytle suggested to her and so, Meg asked, "*Guess what problems your friend asked me yesterday?*" then, dramatically, Gia related quantities such as fish, lions, and rabbits, and made up some problems, though some sentences were not logical nor complete. "*Cats ate many fish, tell me fish...*".

Apart from voicing, they also learned to write what they think. Sharing what they find in the form of what mom named "a diary" is also a successful way to record their problem posing, as in Leung and Wu (2000).

Gia, "*10, 1...fish, 3... friends, 1, how many left?*"

Meg, "*Dear, What do you mean?*"

Gia, "*I said, originally I have ten cats, sent one to the fisherman, then sent one to each of my three friends.*"

When Gia did not express herself clearly and the question "*How many left*" was unclear Meg asked Gia to try telling Mom what the problem was.

There was also evidence of an adult's success in probing a child's thinking and communicative power.

Meg, "Guess what questions I am going to ask you?"

Gia was quiet.

Meg, "*Yesterday I read stories with your friends and asked them to guess. What do you think your friends asked me?*"

Gia, "*How many eyes are there in 5 cats?*"

Meg, "*Any more problems?*"

Gia, "*Yes. They asked about the cats.*"

Meg, "*About the cat, what else?*"

Gia, "*How many claws…*"

Meg, "*How many claws in…*"

Gia, "*6 cats.*"

After talking about cats, Gia also related her experience in going to the zoo, and when Meg invited Gia to make up problems using the zoo context, the response was, "*How many lions are male? How many are female?*" Non-math problems were also present, such as, "*Which families shall I give this cat to*?" "*Why did mom only allow one cat*?" the parent Meg asked, "*If you were the mother, how many cats would you allow?*" Gia just laughed and did not answer. The exchanges of information, problem posing and solving, asking questions, and answering or just talking about something that was not related to the story added warmth to the interactions of a parent and a child. The mothers and their daughters drew on their everyday ways. Meg tried to equalize her position as an adult and even asked the girl to decide how many cats were allowed. The amount of playfulness enabled their mathematical interactions, which was similar to Civil and Quintos (2023).

The inductive and deductive reasoning is a "mathematization" in this problem posing activity. In all, writing about one's thinking is a possible task for young children. Providing an attractive diary sheet or proposing a friend who attempted to guess and share were techniques that motivated children to voice or record their thinking in words or drawings. That some problems are idiosyncratic is also the case in previous studies, even for subjects with older ages (Leung & Silver, 1997; Saenz-Ludlow, 2006). Interestingly, for the story *Being the Fifth*, all the problems posed did not involve more than 5 toys. This limitation may be influenced by the story in this picture book. For the other book, the number of cats was changing and did not extend to more cats than was drawn in the pages of the book. In Leung and Silver (1997), adults were asked to pose math problems using paper-n-pencil tasks, and five types of problems were: *Not-a-problem, Non-Math, Implausible, Insufficient, and, Sufficient).* In this study, with realistic contexts and simple number facts, all were legitimate problems (Sufficient math or non-math problems). There was none on Implausible nor Insufficient.

5 Conclusions and Implications

The investigator concluded that picture books can be a prompt to promote math thinking and problem posing can be exhibited at an early age. Findings in young children suggested a progression in early number learning and in the maturity of problem types in problem posing. Among various instructional strategies for children to pose problems, reading picture books proved to be a feasible means to provoke math thinking and problem posing of young children. When young children read stories with adults, they recall facts from the stories and supply their information to pose problems. Though some problems were not mathematical and some were hard to understand, parents and children in this study enjoyed this process and posed problems that indicated their thinking in early number development.

Conditions for this to be feasible are not limited to the carefully selected tasks, but also to how the parents (adults) talked with the young child. The "how" included adults' strategies in having children recall, re-state, and predict, and techniques such as guessing, and recording problems in diaries, especially when children were in doubt or stayed quiet. Future research can address other problem posing tasks including other techniques in reading with young children, or use other means to promote math thinking and problem posing of this age group; and maybe, also to extend beyond the number of toys (cats) given in the book. The implications are also for enhancing parent–child posing problems as family time at home (see Leung & Wu, 2000).

While math thinking is a priority in math education, how to get it started at a young age is important but challenging for teachers and researchers. As international assessments (e.g. PISA) aim at math literacy for all nations, future research on strategies for young children to think about math and how they pose problems is in need. A possible direction includes cross-cultural differences in young children's math thinking, like the one mentioned in Van den Heuvel-Panhuizen and Van den Boogaard (2008) that compared kindergarten children reading *Being Fifths* in 6 countries. Maybe, problem posing can be an added component to reading picture books. As mentioned, while stories have the potential to prompt children to make utterances and to infer math ideas to pose math problems, future research can attend to study the part on components of problem solving and posing (Fosse et al., 2020).

References

Baroody, A. J. (1983). The development of procedural knowledge: An alternative explanation for chronometric trends of mental arithmetic. *Developmental Review, 3*, 225–230.

Baroody, A. J., Lai, M. L., & Mix, K. S. (2006). The development of young children's number and operation sense and its implications for early childhood education. In B. Spodek & O. Saracho (Eds.), *Handbook of research on the education of young children* (pp. 187–221). Erlbaum.

Baroody, A. J., & Purpura, D. J. (2017). Early number and operations: Whole numbers. In J. Cai (Ed.), *First compendium for research in mathematics education* (pp. 308–354). National Council of. Teachers of Mathematics.

Becker, J. R. (2005). Effects of problem posing instruction. *Proceedings of the 29th Conference of the International Group for the Psychology of Mathematics Education*, Vol. 1, pp. 273. PME.

Charalambos, C., Leonidas, K., & George, P. (2003). *Testing a comprehensive model for measuring problem solving and problem posing skills of primary pupils*. Proceedings of the 27th Conference of the International Group for the Psychology of Mathematics Education, Vol. 2, pp. 205–212. PME.

Civil, M., & Quintos, B. (2023). Mothers and Children doing mathematics together: Implications for teacher learning. *Teachers College Record, 124*(5), 13–29.

English, L. D. (1997). Promoting a problem-posing classroom. *Teaching Children Mathematics, 4*(3), 172.

Fosse, T., Lange, T., & Meaney, T. (2020). Kindergarten teachers' stories about young children's problem posing and problem solving. In M. Carlsen, I. Erfjord, & P. S. Hundeland (Eds.), *Mathematics education in the early years* (pp. 351–368). Springer.

Gelman, R., & Gallistel, C. R. (1978). *The child's understanding of numbers*. Harvard University Press.

Ginsburg, H. P. (1977). *Children's arithmetic*. D. Van Nostrand.

Goldin, G. (2000). A scientific perspective on structured, task-based interviews in mathematics education research. In A. E. Kelly & R. A. Lesh (Eds.), *Handbook of research design in mathematics and science education* (pp. 517–546). Lawrence Erlbaum Associates.

Griffiths, R., & Clyne, M. (1991). The power of story: Its role in learning mathematics. *Mathematics Teaching, 135*, 42–45.

Hasemann, K. (2003). *Anfangsunterricht mathematik*. Spektrum.

Herbel-Eisenmann, B., & Wagner, D. (2010). Appraising lexical bundles in mathematics classroom discourse: Obligation and choice. *Educational Studies in Mathematics, 75*(1), 43–63.

Jandl, E., & Junge, N. (2000). *Vijfde zijn [Being fifth]*. Ploegsma.

Jordan, N. C., Kaplan, D., Locuniak, M. N., & Ramineni, C. (2007). Predicting first-grade math achievement from developmental number sense trajectories. *Learning Disabilities Research & Practice, 22*, 36–46. https://doi.org/10.1111/j.1540-5826.2007.00229.x

Kaufman, E. L., Lord, M. W., Reese, T. W., & Volkmann, J. (1949). The discrimination of visual numbers. *American Journal of Psychology, 62*, 498–525. https://doi.org/10.2307/1418556

Leung, S. S., & Silver, E. A. (1997). The role of task format, mathematics knowledge, and creative thinking on the arithmetic problem posing of prospective elementary school teachers. *Mathematics Education Research Journal, 9*(1), 5–24. Australia.

Leung, S. S., & Wu, R. (2000). Sharing problem posing at home through diary writing: Case of grade one. *The Australian Mathematics Primary Classroom, 5*(1), 28–32.

Leung, S. S. (2023). Parental involvement in young children's mathematical learning: The story of family math tasks. In J. Cai, G. J. Stylianides, & P. A. Kenney (Eds.), *Research studies on learning and teaching of mathematics*. (pp.157–178). Springer.

Marilyn Burns., & Dianne Ochiltree. (2006). *Cats Add Up! Hello reader! Math Scholastic Reader Collection Level 3 Series Scholastic reader.* Scholastic.

Ministry of Education. (2003). *The 9-year alignment curriculum standards in Taiwan*. (In Chinese).

Ministry of Education of Taiwan. (2018). Curriculum Guidelines for the 12-Year Basic Education Elementary School, Junior High School, and Upper Secondary School Domain of Mathematics (in Chinese).

Mulligan, J., & Vergnaud, G. (2006). Research on children's early mathematical development. In Angel Gutierrez & Paolo Boero (Eds.), *Handbook of research on the psychology of mathematics education: Past, present, and future*. Sense Publishers.

National Council of Teachers of Mathematics (NCTM). (2000). *Principles and standards for school mathematics*. Author.

Organization for Economy Co-operation and Development. (2004). *Results of program for international student assessment*. OECD/PISA, Published at 00.01 Paris Time 7 December.

Polya, G. (1945). *How to solve it* (2nd ed.). Doubleday.

Saenz-Ludlow, A. (2006). *A teacher's method to introduce story problems: Student-generated problems.* Proceedings of the 30th Conference of the International Group for the Psychology of Mathematics Education, Vol. 1, pp. 9–16. PME.

Sarama, J., & Clements, D. H. (2009). *Early childhood mathematics education research: Learning trajectories for young children.* Routledge.

Silver, E. A. (1994). On mathematical problem solving. *For the Learning of Mathematics*, 14 (1), 19–28. Plenary lecture in 2003, PME 17, Tsukuba, Japan.

Van den Heuvel-Panhuizen, M., & Van den Boogaard, S. (2008). Picture books as an impetus for kindergartners' mathematical thinking. *Mathematical Thinking and Learning, 10*, 341–373.

Zhang, Q., Sun, J., & Yeung, W.-Y. (2023). Effects of using Picture books in mathematics teaching and learning: A systematic literature review from 2000–2022. *Review of Education, 11*, Article e3383. https://doi.org/10.1002/rev3.3383

Zobairi, N., & Gulley, B. (1989). *The old tale: Oral storytelling and the young child.* (ERIC Document Reproduction Service No. 319517).

A commentary on the parental involvement chapters

Marta Civil

The four chapters centered on parents and mathematics present several efforts that the author designed to engage parents with the ideas behind problem solving. As someone whose research agenda for over 30 years has centered on parental engagement in mathematics, I found these chapters were very interesting to read. It was refreshing to read that as part of the curriculum standards in Taiwan, "schools should encourage parent associations to establish parent-learning communities or parent-teacher co-learning communities" (p. 217, Chapter 11). The idea of parents and teachers learning together and from each other in mathematics can help address the power differential that often exists in school settings, where teachers are viewed as the ones with the knowledge (Ishimaru et al, 2015; Quintos, et al. (2024); Takeuchi, 2018). I like the idea of co-leaning communities, although the rest of excerpt reads "to improve educational knowledge among parents." I understand that this is important particularly in my work with immigrant-origin parents, as they may not be familiar with their children's school system. However, the emphasis should be on this co-learning, in that teachers can also learn a lot from parents' knowledge of mathematics (Varley Gutiérrez et al., 2024); teachers, school personnel, and researchers learning from parents and the community does not happen very often. I argue that this is an key aspect to pursue. In Chapter 12, the Venn diagram on page 5 presents a clear visual of what the author is aiming for with the idea of co-learning, the need to be on the same plane. As she writes, "when there is no sharing of knowledge, the above co-learning model will not work" (p. 239). That there is real sharing where all involved are learning from each other is crucial.

Chapter 11 presents several specific examples of the activities used in the projects that could help other educators adapt them to their settings. In particular, I want to draw attention to the affective discussion in that chapter. The various quotes from parents and students on their appreciation of doing mathematics together are heart-warming. Reading them reminds me of what a mother told me many years ago. She

M. Civil
The University of Arizona, Tucson, USA
e-mail: martac@arizona.edu

had joined the project to work with her middle school daughter (12 years old), to have an opportunity to spend more time with her daughter. She was working long hours and the project gave her a chance to have some dedicated time just with her daughter as she was beginning to navigate adolescence in a different country. While Chapter 11 focuses on younger children, it is worth exploring what the implications are for parent-child interactions in mathematics as the children grow older (Civil & Quintos, 2022; Salazar & Civil, 2024). We see a glimpse of this in Chapter 12, and in Chapter 14 the author in fact describes some of the work in Civil and Quintos (2022).

In the 4th grade classroom case presented in that same Chapter 11, there is some discussion written in terms of parents of high/low achievers. I wonder if we could move away from this labeling of students and focus on the nature on the interactions. While the problems they posed may vary depending on the children's comfort level with mathematics, what can be learned from the experience of children and parents posing problems together? To me an important aspect is what parents reported from the experience, such as spending more time discussing mathematics with their children, becoming co-learners with their children, and doing mathematics together. The quotes in that part of the chapter point to the potential impact on family dynamics, as the child who says that now they were spending more time doing homework together.

The work in Chapter 12 reminds me of the framework we used in MAPPS (Math and Parent Partnerships in the Southwest) (Civil et al., 2005) where we describe the model of parents as parents, as learners, as teachers (facilitators), and as leaders. Chapter 12 presents several examples of parents as learners, teachers, and facilitators. In particular, the section of parents as facilitators illustrates the strategy that the author used to invite parents to come forward as facilitators. The strategy reflects the author's understanding of the cultural norms. This is an important point: key to authentic work with families/parents/caregivers is an understanding of the cultural norms of the community.

Project MAPPS was grounded in prior work with a group of Mexican origin women in Tucson, Arizona, where we developed the concept of parents as intellectual resources. As we write, "key to our approach was researching ways to promote families (parents) as intellectual resources in the academic development of their children" (Civil & Andrade, 2003, p. 153). Our work challenged the traditional view of parental involvement, particularly in minoritized communities, where mothers usually come to school to engage in activities such as "cutting and pasting, monitoring the cafeteria and the playground, ..." (p. 155). Similarly, the author of this book, in Chapter 12, describes the typical roles of parents along similar lines, "arranging books in the library, and helping out with coffee for meetings.... helping with notice boards but very rarely as resources in teaching" (pp. 5–6). For me, it was particularly interesting to see similarities with my work with parents in a different context, Taiwan. It makes me wonder how common this lack of recognition of the potential for parents' intellectual contributions is in schools across the world. This book contributes to challenge this view.

Chapter 13 presents a different view of how parents can engage in mathematical learning with their children through its focus on outdoor activities. The author describes how she used an activity that was already part of the school (Fun Fair) to include mathematics games. The chapter shows the powerful role that parents can play in designing and organizing activities when given the opportunity (see Civil et al., 2023 for an example of a group of mothers designing a games workshop). The author gives examples of parents' input in which problems to include and why. The sampling of problems that the parents posed is very rich and could be very helpful to other educators who want to adapt this project to their setting. As in Chapter 11, this chapter also includes the affective component in parents and children doing mathematics together.

Chapter 14 provides an avenue to engage young children in problem posing through the use of picture books. It includes an overview of the literature on early childhood mathematics, which in itself, independently of the work with parents, can be very useful for educators working at that level.

I have long been interested in connecting the ideas of storytelling, children's books, and mathematics and this chapter addresses this. For example, in our current work we are looking at mother-child interactions and the use of storytelling as way for the mother to support the child's engagement with the problem (Civil & Salazar, 2024). In this chapter, I found the idea of "story-telling mom" quite original and with great potential to engage parents as intellectual resources. As for the books used in the examples shown, while the author describes them as "not written to teach math" (p. 6), I would still describe them as "math books" in that the authors intentionally had mathematics content in them. This is not a critique, but more of an observation, as there are many children's books where the mathematics content is less obvious, which could also lead to problem posing, but perhaps in a different direction. The interactions between Mytle (the author) and the two mothers in the three-person meeting give a glimpse of how Mytle valued the mothers' support of their children's learning through her noticing their strategies with their children.

The four chapters present a rich array of activities and scenarios involving parents and their children doing mathematics together. A strong feature of the work presented in these chapters is that the workshops took place for an extended time. This to me is very important as it allows for the development of rapport among the participants and with the researcher(s) and facilitators. I was curious to know more about who the parents participating in these projects were. Was it mostly mothers? What attracted them to the come? What was their educational background? What were the demographics / SES of the school(s) involved? Overall, these chapters provide educators with multiple suggestions for how to implement programs that engage parents and children in problem posing, while also providing ideas for relevant research related to these programs.

References

Civil, M., & Andrade, R. (2003). Collaborative practice with parents: The role of the researcher as mediator. In A. Peter-Koop, V. Santos-Wagner, C. Breen, & A. Begg (Eds.), *Collaboration in teacher education: Examples from the context of mathematics education* (pp. 153-168). Kluwer.

Civil, M., Bratton, J., & Quintos, B. (2005). Parents and mathematics education in a Latino community: Redefining parental participation. *Multicultural Education*, 13(2), 60–64.

Civil, M. & Quintos, B. (2022). Mothers and children doing mathematics together: Implications for teacher learning. *Teachers College Record*, 124(5), 13–29. https://doi.org/10.1177/01614681221105008.

Civil, M. & Salazar, F. (2024, July). *Culturally responsive mathematics in a bilingual setting: Mothers, teachers, and children together*. Presented at TSG 2.4 - ICME-15, Sydney, Australia. (https://icme15-c10000.eorganiser.com.au/data/clients/1/773/submissions/172677/abstract.pdf).

Civil, M., & Salazar, F., Hernandez, M., Mata, S. & Murdock, L. (2023). Mothers as co-facilitators of mathematics learning experiences. In M. Caspe and R. Hernandez (Eds.), *Reflect, connect, collaborate, lead: Promising practices for partnering with families* (pp. 111–117). IAP.

Ishimaru, A. M., Barajas-López, F., & Bang, M. (2015). Centering family knowledge to develop children's empowered mathematics identities. Journal of Family Diversity in Education, 1(4), 1–21.

Quintos, B., Turner, E. & Civil, M. (2024). Parents and teachers collaborating to disrupt asymmetrical power positions in mathematics education. *ZDM-Mathematics Education*. https://doi.org/10.1007/s11858-024-01555-1.

Salazar, F. & Civil, M. (2024). Parent-child interactions: Mothers doing mathematics with their children. In T. Evans, O. Marmur, J. Hunter, G. Leach, & J. Jhagroo (Eds.). *Proceedings of the 47th Conference of the International Group for the Psychology of Mathematics Education* (Vol. 1, p. 207). PME.

Takeuchi, M. A. (2018). Power and identity in immigrant parents' involvement in early years mathematics learning. Educational Studies in Mathematics, 97(1), 39–53. https://doi.org/10.1007/s10649-017-9781-4.

Varley Gutiérrez, M., Napp-Avelli, C., Quintos, B., Salazar, F., Turner, E., & Civil, M. (2024). Families and teachers doing mathematics tasks together: Considering funds of knowledge and positioning. *Mathematics Teacher Educator*, 13(1), 282–296. https://doi.org/10.5951/MTE.2023-0026.

Index

S. S. Leung, *Mathematics Problem Posing in Action*,
https://doi.org/10.1007/978-981-96-8480-9

FSC
www.fsc.org

MIX
Papier aus verantwortungsvollen Quellen
Paper from responsible sources
FSC® C105338

If you have any concerns about our products,
you can contact us on
ProductSafety@springernature.com

In case Publisher is established outside the EU,
the EU authorized representative is:
Springer Nature Customer Service Center GmbH
Europaplatz 3, 69115 Heidelberg, Germany

Printed by Libri Plureos GmbH
in Hamburg, Germany